Time Series and Networks Analysis

Theodoros Karakasidis • Avraam Charakopoulos

Time Series and Networks Analysis

A hands-on approach (Matlab & Octave)

Springer

Theodoros Karakasidis
Department of Physics
Faculty of Science
University of Thessaly
Lamia, Greece

Avraam Charakopoulos
Department of Civil Engineering
School of Engineering
University of Thessaly
Volos, Greece

ISBN 978-3-031-92630-3 ISBN 978-3-031-92628-0 (eBook)
https://doi.org/10.1007/978-3-031-92628-0

This Springer imprint is published by the registered company Springer Nature Switzerland AG
The registered company address is: Gewerbestrasse 11, 6330 Cham, Switzerland

To our families
To our students

Preface

Time series analysis is a dynamic research area that has fundamental importance for a wide range of scientific fields, in terms of both a fundamental point of view and applications ranging from physics to engineering as well as biomedical and financial applications. Over the last few decades, many computational methods have been applied and invented to deal with time series-related problems.

Beyond conventional time series methodologies, this book introduces the analysis of time series through the perspective of complex networks. By transforming time series data into network, researchers can extract topological features that reveal underlying patterns.

But what is extremely interesting in this book is that we also provide detailed guidelines about the software with examples in MATLAB so that anyone can run the corresponding routines and apply them to the data provided or their personal data of research. Moreover, the readers can modify the routines appropriately in order to add more characteristics to the output. This fact is based on the basic idea of learning by trying.

The aim of this book is to provide basic knowledge of time series, introduce some statistical tools useful for analyzing these series, and gain experience in applying various linear, nonlinear, and advanced methodologies. We begin with the basic concepts of asset returns and a brief introduction to the processes to be discussed throughout the book.

The structure of the book is divided into two main parts. Part I: Linear & Non-Linear Analysis introduces the reader to fundamental statistical tools and methods used in traditional time series analysis. It begins with Chap. 1, which focuses on basic statistical analysis, including descriptive measures, distributions, and hypothesis testing for time series data. Chapter 2 delves into the temporal characteristics of time series, exploring properties such as stationarity, autocorrelation, and trends. Following that, Chap. 3 addresses nonlinear dynamics, including phase space reconstruction, chaos indicators, and methods for detecting complex, nonlinear behavior in time-dependent data. Part II: Complex Network Analysis presents a more advanced perspective, applying network science techniques to the study of time series. In Chap. 4, readers are introduced to complex network representations of time series, such as visibility graphs, and how these can be used to analyze structure, connectivity, and patterns in the data. Finally, Chap. 5 includes extended case studies that combine both traditional and network-based methods, showcasing their application to real-world datasets across different scientific disciplines. This structure allows the reader to progress from foundational concepts to more sophisticated analytical techniques, building a comprehensive understanding of time series and their underlying dynamics.

The book can be employed in undergraduate courses as well as in graduate courses, particularly those aiming to provide a fast and practical guide for performing analysis of time series for research purposes in disciplines such as physics, materials science, engineering, and finance, to mention a few areas.

Lamia, Greece — Theodoros Karakasidis
Volos, Greece — Avraam Charakopoulos

Competing Interests The authors have no competing interests to declare that are relevant to the content of this manuscript.

Contents

Part I
Linear & Non-Linear Analysis

Chapter 1
Time Series Statistical Analysis

1.1 Introduction to Time Series

Time series data consists of a set of values that are assembled over even intervals in time and ordered in a chronological order. The time interval at which data is collected is commonly referred to as the time series sampling rate Δt and the inverse of that is the sapling frequency Fs = $1/\Delta t$, i.e., how many points are recorded in a time unit [3, 4].

A time series is a sequential data set of points denoted as $x(t)$, $t = 0, 1, 2, \ldots.$ where x is the variable and t represents time and is defined as follows

$$X(t) = x_1,\ x_2,\ x_3, \ldots x_t$$

The variable we study can be discrete or continuous. Discrete variables can take only given values, i.e., the number of times it rained during a given period, while continuous variables can take any value in a given interval, i.e., the temperature can be 23.4, 23.5, etc [5, 6].

In general, the data are recorded at equal time intervals, although this is not necessarily always the case. In the present book, we are going to deal with time series of the former type. The interest of studying time series is that since they originate from a given system, they can provide information about the underlying system which quite often is too complex to be studied considering all its aspects. A given system can result in several time series as output, i.e., when we study the weather in a given place, we measure the temperature, humidity, atmospheric pressure, etc [8]. Thus, time series analysis can deal with just one time series at a time; in this case, we speak about *univariate* time series analysis. However, one can see the combined analysis of two or more time series simultaneously speaking in this case for *bivariate* or *multivariate* analysis [9, 10].

Time series can originate from various sources. One of the most common sources is computational models of systems, such as simple models or complex simulations

T. Karakasidis, A. Charakopoulos, *Time Series and Networks Analysis*,
https://doi.org/10.1007/978-3-031-92628-0_1

like computational fluid dynamics simulations. The other source originates from sensors used in laboratory experiments and sensors deployed in the field, such as environmental measurements. Financial interest time series, either of the stock exchange or other types, are also of particular interest. Below, we present some examples of such time series.

The data and the codes employed in the present book can be downloaded from the website https://github.com/avcharak/Time-Series-and-Networks-Analysis.

1.1.1 Examples of Time Series

In the following, we present representative examples of time series, some of which will be used as case studies in forthcoming chapters. Indicatively, these examples cover some of the major scientific areas such as economic time series, environmental, applied science, physical sciences but also, we first present simple synthetic time series.

What To Do First Although it may appear silly, our first job when studying a given dataset is to plot the observations as a function of the time. This permits us to see if there is any anomaly in the data, any missing points, or if everything has passed smoothly and with precision from the source file to the data analysis software. Sometimes some commas or points corresponding to different measuring systems can cause headaches. Also, data from a malfunctioning instrument can have data that are not correct.

Simple Numeric Time Series

Here are some examples of simple synthetic time series. The equations of the program are initially given for the purpose of expression, and then the graph, which is the result, is presented.

Example 1 Generation of periodic time series (sine wave) with input of variables, sampling period, frequency, and amplitude. We plot a periodic function using the script 1.1. As the first important step in time series analysis is to visualize the data, execute the script 1.1 in the Matlab command window. The result is displayed in Fig. 1.1.

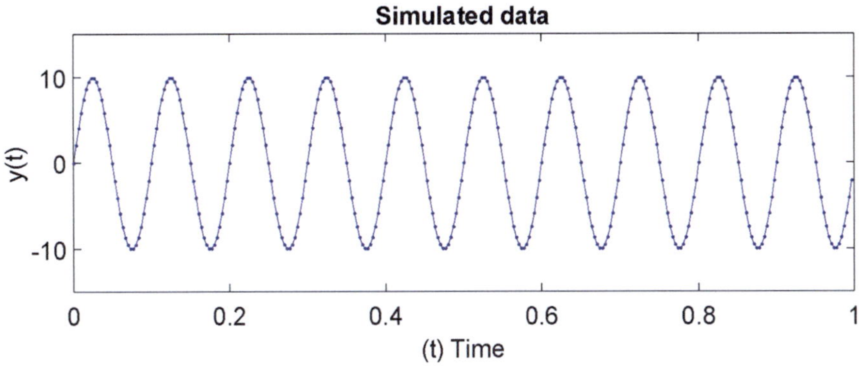

Fig. 1.1 Sinus time series with sampling period Fs = 300, frequency f1 = 10, and amplitude amp = 10

```
%Script 1.1
%Example 1, Generate a periodic time series

Fs=input('Give the sampling period_:');  %sampling period (samples
per second)
f1=input('Give the frequency_:');
Amp=input('Give the Amplitude_:');

Ts=1/Fs;         % seconds per sample
dt=0:Ts:1-Ts;    %signal duration

[y1]=(Amp*sin(2*pi*f1*dt))'; % sin function

plot(dt,y1,'b.-');    %plot function
ylim([-Amp-5 Amp+5])
ax=gca;              %grid parameter
ax.YGrid= 'on';      % y grid on

title('Simulated data','FontSize',20)
ylabel('y(t)')
xlabel('(t) Time')
```

As expected, the data show periodic behavior commensurate with the input variables; otherwise, the time series has periodicity or seasonality.

There are several ways to plot time series on Matlab. Although the signal may appear continuous, it is actually a discrete signal since there is a large number of data points. Lines are generally used to guide our eyes. Sometimes, the number of points can be so large that the points appear to form a continuous line.

Example 2 We can also create more complex synthetic signals by combining simple signals. For instance, we can create three periodic time series with different frequencies by running in the command window and summing them using script 1.2, and the results appear in Fig. 1.2. For this complex time series, we can say that

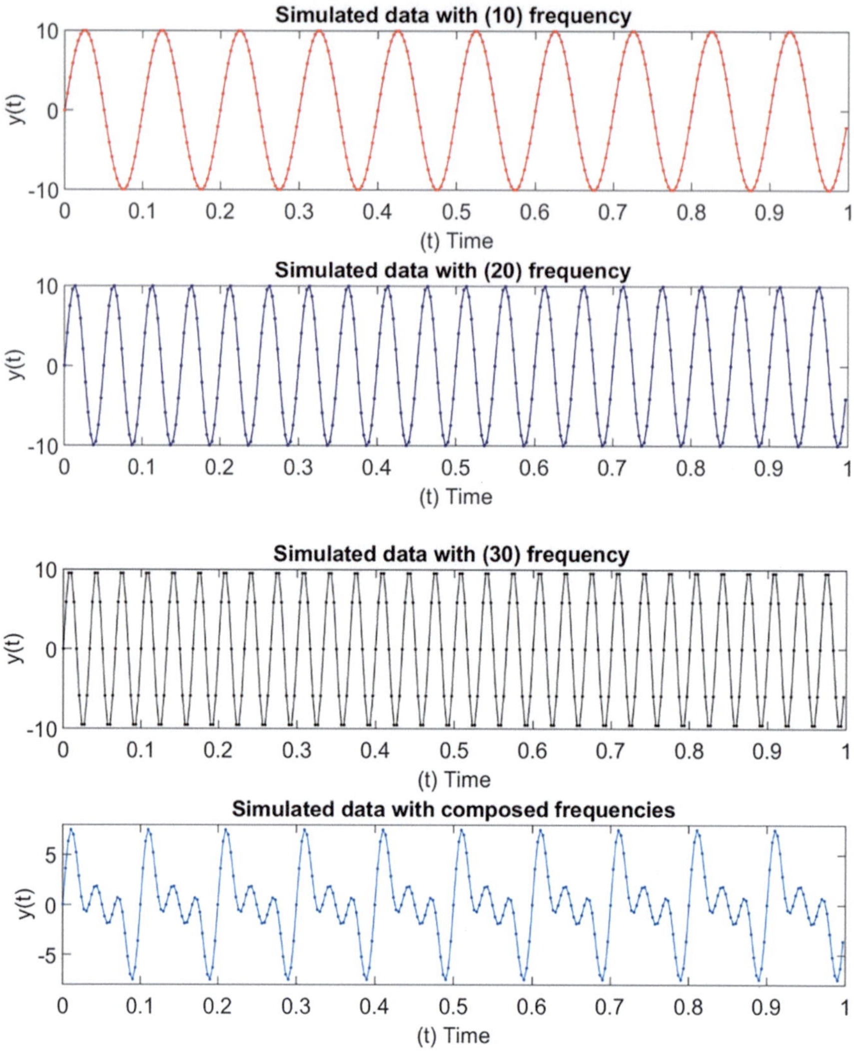

Fig. 1.2 Generate three time series with different frequencies (sampling frequency Fs = 300, amplitude amp = 10, and signal frequencies f1 = 10 Hz, f2 = 20 Hz, f3 = 30 Hz), respectively, in one plot

it still has periodicity, which depends on the frequencies of the individual time series from which it is composed.

```
%Script 1.2
%Example 2, Generate a synthetic periodic time series

Fs=input('sampling frequency_');
f1=input('frequency_1_');
f2=input('frequency_2_');
f3=input('frequency_3_');
Amp=input('Amplitude_');

Ts=1/Fs;          %sampling period
dt=0:Ts:1-Ts;     %signal duration

y1=Amp*sin(2*pi*f1*dt);
y2=Amp*sin(2*pi*f2*dt);
y3=Amp*sin(2*pi*f3*dt);
y4=0.3*(y1+y2+y3);

subplot(4,1,1);
plot(dt,y1,'r.-','MarkerSize',10);
title(sprintf('Simulated data with (%d) frequency
',f1),'FontSize',18)
ylabel('y(t)')
xlabel('(t) Time')
subplot(4,1,2);
plot(dt,y2,'b.-','MarkerSize',10);
title(sprintf('Simulated data with (%d) frequency
',f2),'FontSize',18)
ylabel('y(t)')
xlabel('(t) Time')
subplot(4,1,3);
plot(dt,y3,'k.-','MarkerSize',10);
title(sprintf('Simulated data with (%d) frequency
',f3),'FontSize',18)
ylabel('y(t)')
xlabel('(t) Time')
subplot(4,1,4);
plot(dt,y4,'.-','MarkerSize',10);
title('Simulated data with combined frequencies','FontSize',18)
ylabel('y(t)')
xlabel('(t) Time')
```

Field Measurement Data: Environmental Time Series

The next series contains 30 years of weekly observations recorded by a meteorological station, measuring the horizontal wind speed as a weekly mean across all directions. This dataset comprises a total of 1488 records [7]. In Fig. 1.3, the weekly mean wind speed vs time is presented.

By executing script 1.3 in Matlab command window, we get the time series plot.

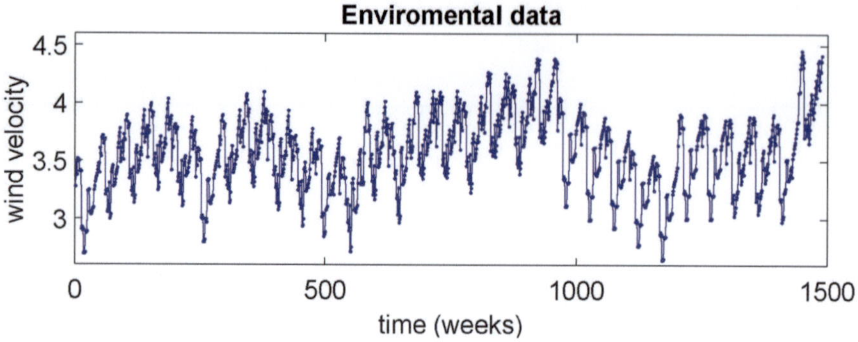

Fig. 1.3 Weakly mean wind speed (http://www.emy.gr/emy/el/)

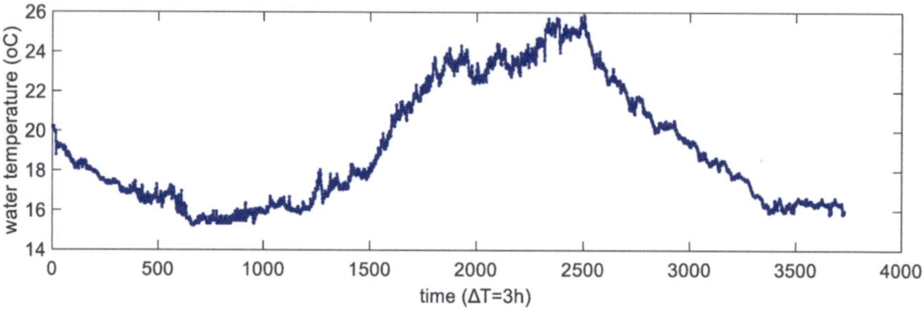

Fig. 1.4 Water temperature 3-hourly data measured (https://poseidon.hcmr.gr/)

```
%Script 1.3
%Example 3, plot wind time series

plot(wind,'b.-','MarkerSize',10);    %'wind' the name of time
series

title('Environmental data','FontSize',20)
ylabel('wind velocity','FontSize',10)
xlabel('time (weeks)','FontSize',10)
```

By a first look in Fig. 1.3, we can understand that several smaller and larger periodicities are present in this phenomenon. We will describe later in the book methodologies in order to derive the corresponding frequencies in the time series since they may reveal information about the system under study.

Another example of a time series of environmental interest is presented in Fig. 1.4 (after running script 1.4), where we can see seawater temperature data recorded every 3 h through a system of buoys for measuring atmospheric and oceanographic data (https://poseidon.hcmr.gr/) [1].

```
%Script 1.4
%Example 4, plot water temperature time series

plot(water_temp,'b.-','MarkerSize',10);   %water_temp the name of
time series
title('Enviromental data','FontSize',20)
ylabel('water temperature (oC)','FontSize',10)
xlabel('time (ΔT=3h)','FontSize',10)
```

Experimental Data: Time Series from Applied Sciences

The following time series examples come from the scientific field of applied sciences. After running script 1.5, we can see in Fig. 1.5 the instantaneous temperature recorded during an experiment of a vertical turbulent heated jet [2]. It is clear how complex the time series behavior is compared to the previous examples.

```
%Script 1.5
%Example 5, plot temperature time series

plot(turb_jet,'b.-','MarkerSize',6);   %Turb_jet the name of time
series
title('Enviromental data','FontSize',20)
ylabel('temperature (oC)','FontSize',10)
xlabel('time (sec)','FontSize',10)
```

Financial Time Series

The next series we consider here is the daily index of the Nasdaq stock market index and gold price for 1 year (the data can be downloaded from Yahoo Finance). Using script 1.6, we plot the data against the time index, and the results are presented in Fig. 1.6. We can see that these time series present a different behavior from previous ones.

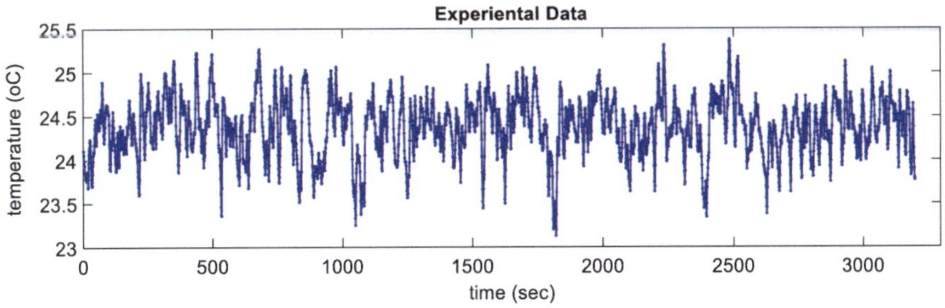

Fig. 1.5 Temperature time series of experimental temperature time series from a vertical turbulent heated jet [2]

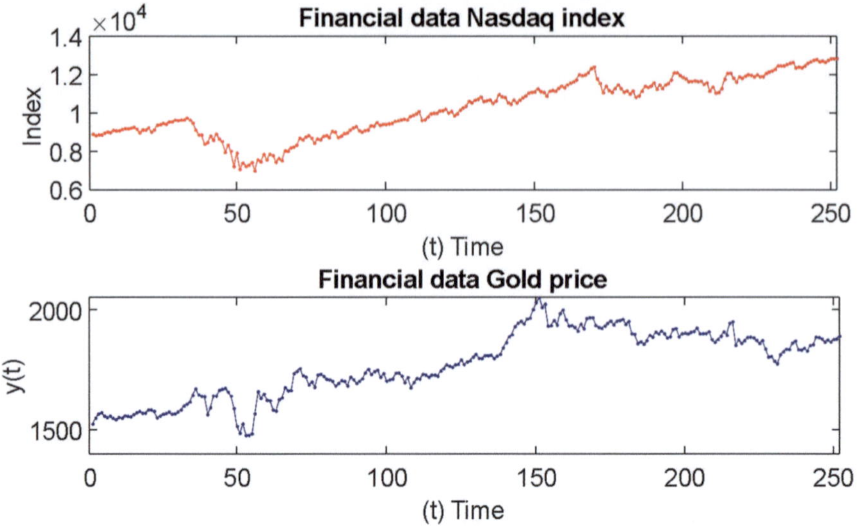

Fig. 1.6 Nasdaq 100 index and gold price for the same time period (data from Yahoo Finance)

```
%Script 1.6
%Example, plot financial time series

figure
subplot(2,1,1);
plot(Nasdaq,'r.-','MarkerSize',10);
xlim([0 252])
title('Financial data Nasdaq index','FontSize',18)
ylabel('Index ')
xlabel('(t) Time')
subplot(2,1,2);
plot(gold,'b.-','MarkerSize',10);
xlim([0 252])
title('Financial data Gold price','FontSize',18)
ylabel('y(t)')
xlabel('(t) Time')
```

1.2 Statistical Analysis: Univariate, Bivariate, and Multivariate

In statistical analysis, there exists, among other things, three main categories of analysis, depending on the origin and the number of variables: univariate, bivariate, and multivariate analysis. When the measurement data refers to a sequence of measurements of the same variable collected over time, then we have the case of the univariate analysis.

Bivariate data involves two different variables and mainly, the analysis of these data focuses on investigating the causes and relationship between these two variables, while multivariate analysis is a more complex form of statistical analysis technique and is used when there are more than two variables in the data set. Multivariate analysis allows the separate and combined effects of the independent variable to be examined.

1.3 Descriptive Statistics

Techniques used to summarize and describe the characteristics of a group or to compare characteristics between groups are known as descriptive statistics.

1.3.1 Mean, Median, Variance, Standard Deviation, Max and Min, Histogram, Skewness, and Kurtosis

Below we provide the definition of the most common descriptive statistics measures. In general, when we study a problem, this corresponds to a population; for example, all temperatures in a class of experiments. However, we have access only to a limited number of measurements, thus we have access to a sample of the population. There are whole books dealing with the ways to perform appropriate sampling, and thus we are not going to treat this subject here. However, below we are going to treat terms that often appear as population or sample properties.

The **sample average or mean**, $\underline{x}$, is obtained by summing up all measurements and dividing by their number N:

$$\underline{x} = \frac{x_1 + x_2 + \ldots + x_N}{N} = \frac{1}{N}\sum_{i=1}^{N} x_i \tag{1.1}$$

The mean represents the value around which the observations are gathered. When referring to the ideal population average quite often, the symbol μ is used.

The **median** is the middle value in a group of numbers ranked by value. It is the number that is exactly in the middle, so 50% of the ranked numbers are above and 50% are below the median. This definition holds when the number of observations is odd. If the number of observations is even, then we take the average of the two middle values as a median.

The **min** is simply the lowest value of the sample, while the **max** is the highest value.

The **sample variance (σ^2)** is the sum of the difference of each point from the mean. As it is known, it is a measure of spread of the values around the mean:

$$\sigma^2 = \frac{1}{\mathrm{N}-1}\sum_{i=1}^{\mathrm{N}} \left(x_i - \underline{x}\right)^2 \tag{1.2}$$

σ is called the standard deviation. When referring to population variance and standard deviation, the symbols S^2 and S are used, respectively.

The **mode** is the value that appears at a higher frequency (i.e., more frequently).

Range is the difference between the dataset's largest value and the dataset's smaller value:

$$R = x_{\max} - x_{\min}$$

There are also other measures known as shape measures, which help describe how data points in a dataset are distributed around the mean. These measures assist in identifying patterns that become evident when the data is visualized on a graph.

Histogram It is of interest to measure the frequency of appearance of observed values. Since, in many cases, we have measures that can take continuous values in fact what we do is that we divide the region of values into intervals, and we measure the values that fall within the given intervals. This leads to the so-called histogram of the data. This representation can provide quite an insight into how the values are distributed around the mean.

In Fig. 1.7, we present the histogram for the time series of wind speed using script 1.7. The blue bars show the frequency of the value where the red line represents the normal distribution curve. We can see that in this case, we have a pretty symmetric distribution of values around the mean value of the measurements. We also can see the width of the distribution of values around the mean.

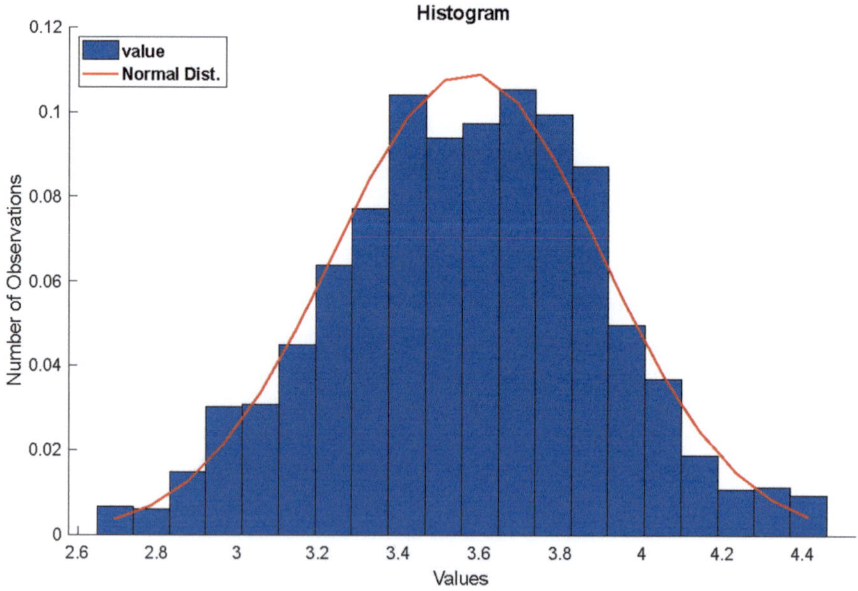

Fig. 1.7 Histogram of wind speed time series along with a fit corresponding to a normal distribution

```
%Script 1.7
%Example, calculates the histogram of wind time series

figure
function HIST=histogram_avra(data)

close all
m=length(data);
mu=mean(data);
sigma=std(data);
Bins=20;
[freq,x_axis]=hist(data,Bins);
pdf=freq./m;
figure('position',[0 0,800 600]);
title('Histogram')
xlabel('Values');
ylabel('number of Observations');
hold on
Y=normpdf(x_axis,mu,sigma);
bar(x_axis,pdf,'FaceColor','blue','BarWidth',1);
hold on
plot(x_axis,Y./sum(Y),'Color','red','LineWidth',2);
hold on
       if nargin>=2
       m_conf=length(percentiles);
       quant=quantile(data,percentiles);
       X_axis=get(gca,'XTick');
       set(gca,'XTick',unique([min(X_axis) mu quant
max(X_axis)]));
       for i=1:m_conf
           plot(quant(i),0,'sm','LineWidth',2);
           hold on
       end
       legend({'value','Normal
Dist.','Mean','Quantile'},'Location','NorthWest','FontSize',14,'Fo
ntWeight','Bold');
    else
       quant='';
       legend({'value','Normal
Dist.','Mean'},'Location','NorthWest','FontSize',14,'FontWeight','
Bold');
    end
```

In the following, we will present some brief information about various known distributions that occur in many study systems. The main characteristic of these distributions is the shape of the corresponding histograms generated from the recorded values when examining the related systems.

Normal Distribution

Gauss introduced the normal distribution or Gauss distribution, and the probability of a value x occurs has the form:

$$f(x) = \frac{1}{\sqrt{2\pi\sigma}} e^{-\frac{(X-\mu)^2}{2\sigma^2}} \quad -\infty < X < +\infty \tag{1.3}$$

We follow the notation for the X variable $X \sim N(\mu, \sigma^2)$ where μ is the population average and σ^2 is the population variance. In the case of the normal distribution, we have the following percentages of the population in the corresponding intervals:

$$\begin{aligned} &[\mu - \sigma, \mu + \sigma] && 68,27\% \text{of the population} \\ &[\mu - 2\sigma, \mu + 2\sigma] && 95,45\% \text{of the population} \\ &[\mu - 3\sigma, \mu + 3\sigma] && 99,73\% \text{of the population.} \end{aligned}$$

The above is approximately valid for a sample with a mean x and variance s and is represented schematically in Fig. 1.8.

What is of interest is the so-called standard normal distribution:

$$f(x) = \frac{1}{\sqrt{2\pi}} e^{-\frac{Z^2}{2}} \quad -\infty < X < +\infty \tag{1.4}$$

which corresponds to $\mu = 0$ and $s = 1$ and is obtained if we transform the variable X using the relation:

$$Z = \frac{X - \mu}{\sigma} \tag{1.5}$$

The Z variable is known as the standardization of X and is often also known as the Z-score. This is very practical when we have to compare different sets of values with different mean values and variances. Of course, we use the corresponding sample quantities in the case of a sample.

We must mention here that other well-known distributions can also be applied in exceptional cases. Such distributions are the gamma distribution and Weibull distribution, which we discuss below.

There are two measures of form that describe the shape of the distribution and are discussed below.

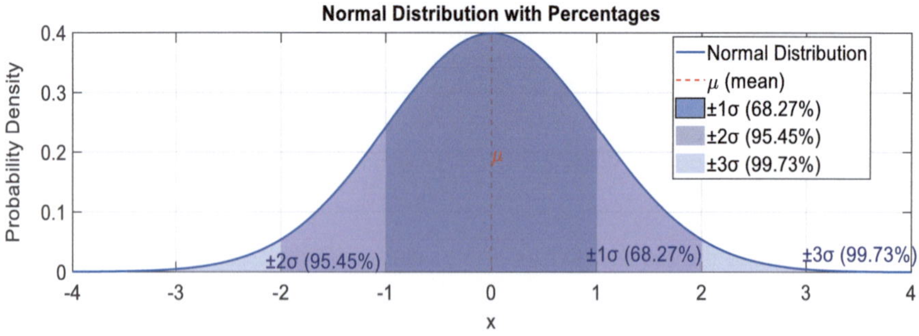

Fig. 1.8 Normal distribution of time series

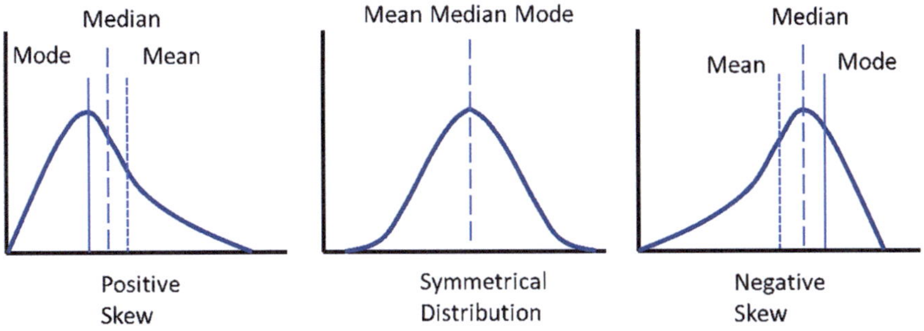

Fig. 1.9 Probability distribution for different situations

Pearson's moment coefficient of skewness is the the average of the standardized cubed deviation from the mean $(\overline{x})$:

$$b1 = \frac{1}{\mathrm{N}-1}\frac{\sum_{i=1}^{\mathrm{N}}(x_i-\overline{x})^3}{S^3} \tag{1.6}$$

- If $b1 = 0$, the distribution is symmetric (Fig. 1.9 center),
- If $b1 < 0$, the distribution presents negative asymmetry (skewness) (Fig. 1.9 right).
- If $b1 > 0$, the distribution presents positive asymmetry (skewness) (Fig. 1.9 left).

The kurtosis ($b2$) sample coefficient is the average of the fourth power of the standardized deviations from the mean:

$$b2 = \frac{1}{\mathrm{N}-1}\frac{\sum_{i=1}^{\mathrm{N}}(x_i-\overline{x})^4}{S^4} - 3 \tag{1.7}$$

- If $b2 = 0$, the distribution is called Mesokurtic.
- If $b2 < 0$, the distribution is characterized as Leptokurtic.
- If $b2 > 0$, the distribution is called Platykurtic.

These different situations are presented schematically in Fig. 1.10.

The simplest time series model is the one that has no trend or seasonality, and the variables are independent of each other and are randomly distributed with zero mean:

$$X_t \sim N(0,\sigma^2) \tag{1.8}$$

Thus, we generate a random array of data that follows the normal or Gaussian distribution $N(\mu,\sigma^2)$. This time series has mean $\mu = 0$ and standard deviation equal to one $\sigma = 1$, i.e., the normal distribution $N(0,1)$. Using script 1.8, we produce such a time series, which is plotted in Fig. 1.11.

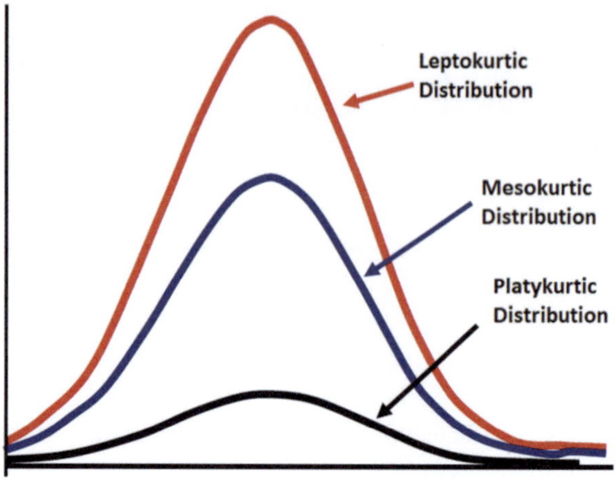

Fig. 1.10 The three characteristic types of distributions are based on kurtosis

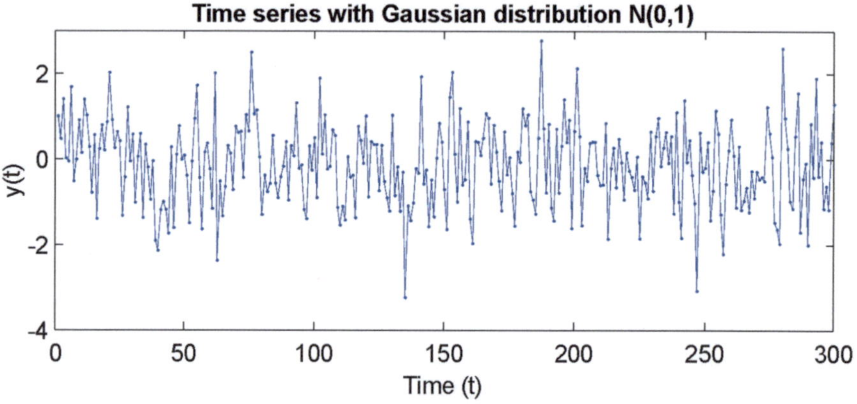

Fig. 1.11 Time series following Gaussian distribution

```
% Script 1.8
% Generate time series with Gaussian distribution

N=input('Give the time series length_:');

TS=randn(N,1);
plot(TS,'.-')
title('Time series with Gaussian distribution
N(0,1)','FontSize',14)
xlabel('Time (t)');
ylabel('y(t)');
```

In order to calculate the main descriptive statistics of the corresponding time series, we can execute script 1.9, and in Fig. 1.12, we can see the time series with

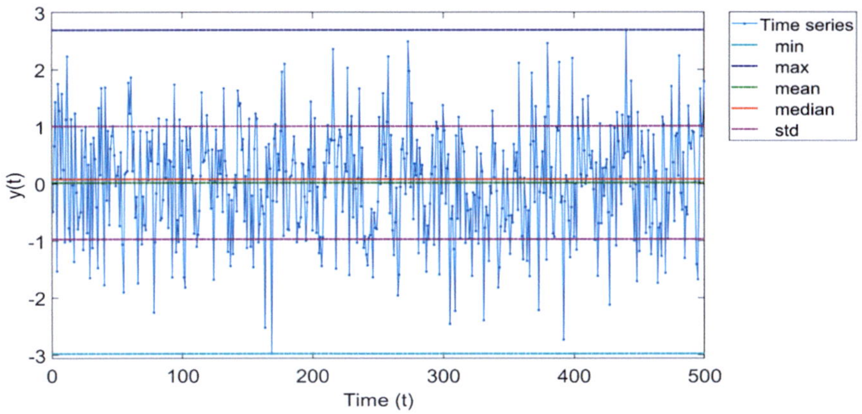

Fig. 1.12 Time series marked the main statistical properties

Table 1.1 Descriptive statistics for time series of Fig. 1.12

Mean	Standard deviation	Variance	Median	Maximum value	Minimum value
0.0183	0.9886	0.9973	0.0774	2.6929	−2.9785

the graphical representation of the corresponding statistical measures, which are presented in Table 1.1.

```
%Script 1.9
%Descriptive statistics

TS=input('Give the time series_:');

Descriptive_results=[mean(TS)' std(TS)' var(TS)' median(TS)'
max(TS)' min(TS)']
plot(TS,'b.-','MarkerSize',10)
xlabel('Time (t)');
ylabel('y(t)');
```

In Fig. 1.13, we present the histogram of the $N(0,1)$ distribution by executing script 1.10.

```
% Script 1.10
% Histogram

TS=input('Give the time series_:');
histfit(TS',25,'normal')
title('Gaussian distribution N(0,1)','FontSize',14)
legend('Data distribution','Normal distribution')
xlabel('y(t)');
ylabel('Frequency');
```

As expected, the distribution of the simulated data is quite close to the ideal normal distribution. Slight differences are due to the finite number of data and the size of the value intervals.

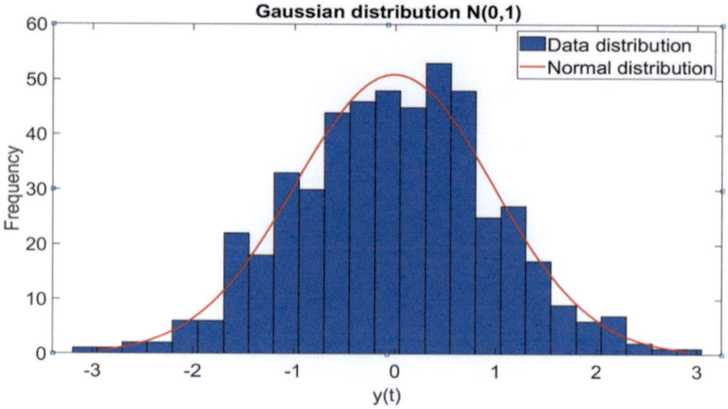

Fig. 1.13 Histogram of the time series $N(0,1)$

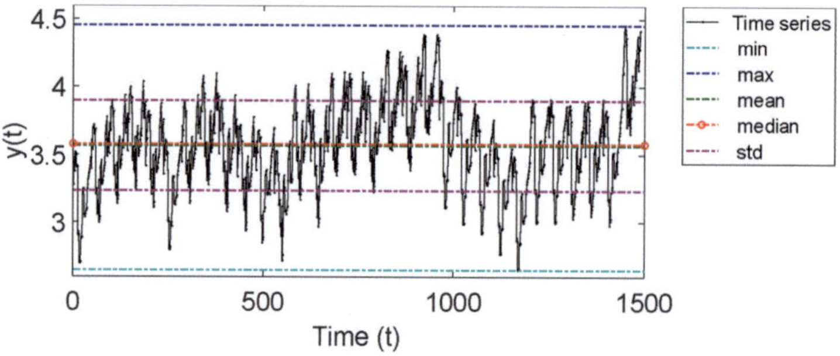

Fig. 1.14 Time series where the main statistical properties are marked graphically on the plot

Table 1.2 Descriptive statistic

Mean	Std	Var	Skew	Kurtosis	Median	Max	Min
3.5690	0.3325	0.1105	−0.0717	2.7957	3.5800	4.4600	2.6500

In the following, we present results of the descriptive statistics of the example 1.2 (Fig. 1.14, Table 1.2).

We can also execute the script 1.11 to calculate the statistics of water temperature time series, and in Fig. 1.15 we present the corresponding histogram of wind time series by executing the script 1.10.

```
%Script 1.11
%Descriptive statistics 2

TS=input('Give the time series_:');

Descriptive_results=[mean(TS)' std(TS)' var(TS)' skewness(TS)'
kurtosis(TS)' median(TS)' max(TS)' min(TS)']
plot(TS,'b.-','MarkerSize',10)
xlabel('Time (t)');
ylabel('y(t)');
```

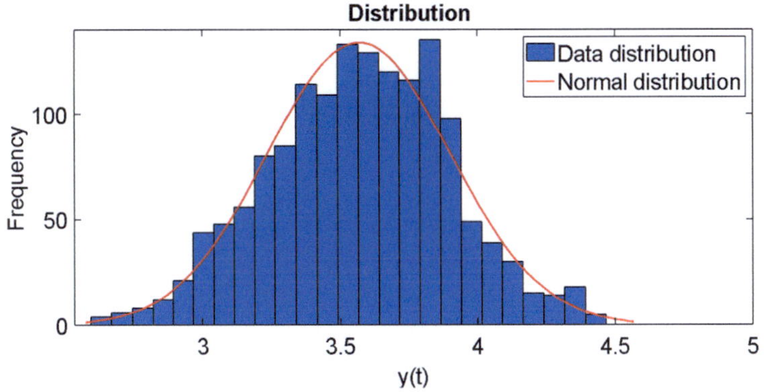

Fig. 1.15 Histogram of time series of wind speed

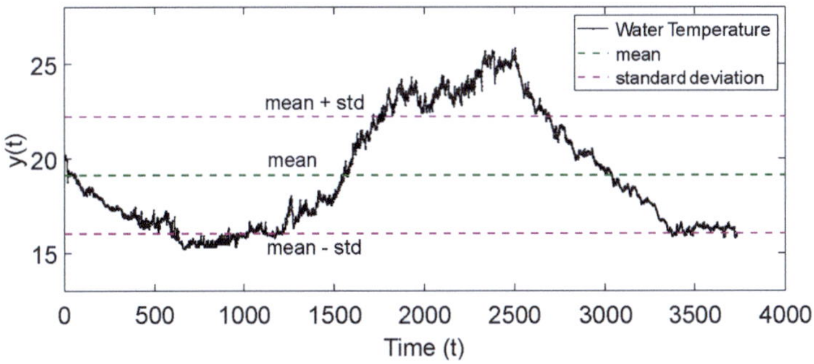

Fig. 1.16 Time series showing the main statistical properties

Table 1.3 Descriptive statistic

Mean	Std	Var	Skew	Kurtosis	Median	Max	Min
19.1458	3.0892	9.5433	0.5638	1.8845	18.0371	25.8252	15.2295

We can see that the histogram is quite close to the normal distribution (represented by the red curve).

We get the following results by executing script 1.9 on the water temperature data (Fig. 1.16, Table 1.3).

We apply the same procedure in the case of the gold price for the period (see Fig. 1.17, Table 1.4) and the Nasdaq index (see Fig. 1.18, Table 1.5).

The time series we have seen so far have practically zero skewness. Below we present some time series that have either positive or negative skewness. This property of time series is well reflected in the histograms where we can clearly see that value distribution differs from the normal distribution and in fact belongs to different kind of probability distributions (thus presenting different histograms too).

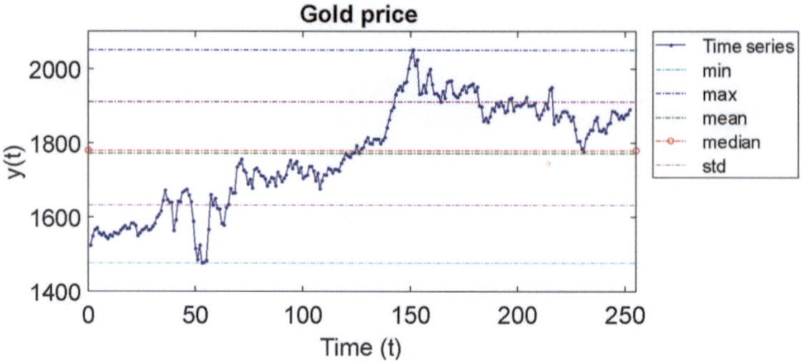

Fig. 1.17 Time series showing the main statistical properties (data from Yahoo Finance)

Table 1.4 Descriptive statistic

Mean	Std	Var	Skew	Kurtosis	Median	Max	min
1773	1400	1952.3	0	0.2	1780	2051	1477

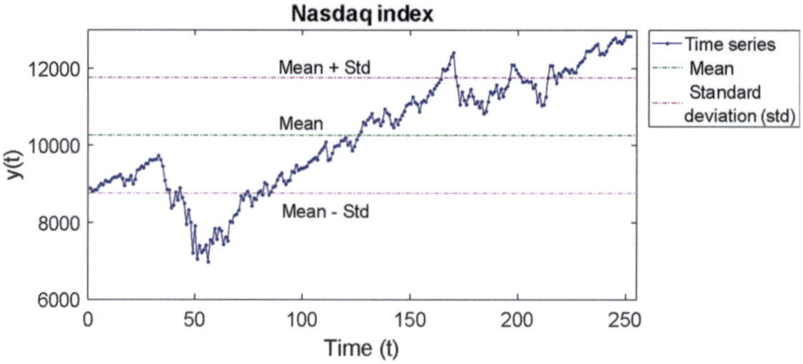

Fig. 1.18 Time series of Nasdaq index showing the main statistical properties (data from Yahoo Finance)

Table 1.5 Descriptive statistic

Mean	Std	Var	Skew	Kurtosis	Median	Max	min
10,300	1500	2,249,000	0.00	0.00	10,300	12,800	7000

Gamma Function

There are time series which, due to the nature of the system from which they originate, present positive or negative skewness value and thus correspond to types of distribution that differ from the normal distribution. A well-known such distribution is the gamma function. The probability density function of the gamma distribution is defined by:

$$f(x)=\frac{\left(\frac{x-\mu}{\beta}\right)^{\gamma-1} * \exp\left(-\frac{x-\mu}{\beta}\right)}{\beta\Gamma(\gamma)} x \geq \mu;\gamma,\beta>0 \tag{1.9}$$

where γ is the shape parameter, μ is the location parameter, β is the scale parameter, and Γ is the gamma function which is described by the following relation:

$$\Gamma(\alpha)=\int_0^\infty t^{a-1}e^{-t}\mathrm{d}t \tag{1.10}$$

Script 1.12 that follows let us generate such a time series graphically represented in Fig. 1.19.

```
% Script 1.12
% Generates gamma random Numbers

r2=gamrnd(3,0.1,1,1000);

plot(r2,'.-')
xlabel('Time (t)');
ylabel('y(t)');
legend('Simulated time series','Location','Best')
```

The results from the calculation of the descriptive statistical measures by executing script 1.11 are presented in Table 1.6.

As observed this time series presents a positive skewness. In statistics, a positively skewed (or right-skewed) distribution is characterized by most values clustering around the left tail of the distribution, while the right tail of the distribution is longer.

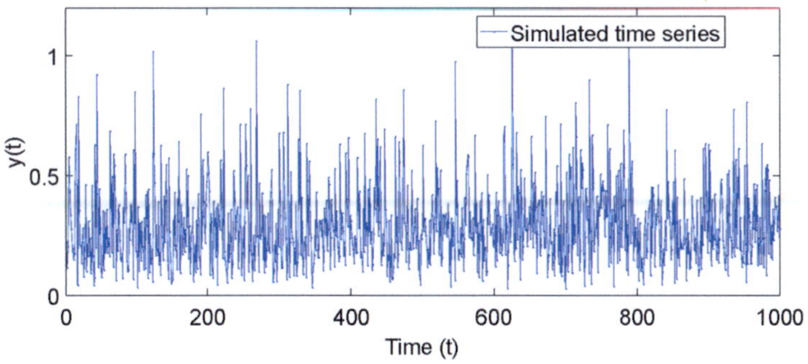

Fig. 1.19 Simulated time series following the Gamma distribution

Table 1.6 Descriptive statistics

Mean	Std	Var	Skew	Kurtosis	Median	Max	min
0.292	0.173	0.0301	1.240	5.116	0.259	1.131	0.028

```
%Script 1.13
%Histogram of time series with positive skewness

TS=input('Give the time series_:'); % i.e r2
Skewness = input('Give the skewness_:'); % i.e 1.24 skewness of r2
histfit(TS',25,'gamma')
xlabel('y(t)');
ylabel('Frequency');
legend(sprintf('Skewness(%d)',Skewness),'Location','Best')
```

By running in the command window script file 1.13, we plot the distribution in Fig. 1.20.

To understand the difference with the normal distribution, we can add the normal distribution curve to the figure by executing the following command.

```
hold on
histfit(r2',25,'normal')
```

The results are depicted in Fig. 1.21.

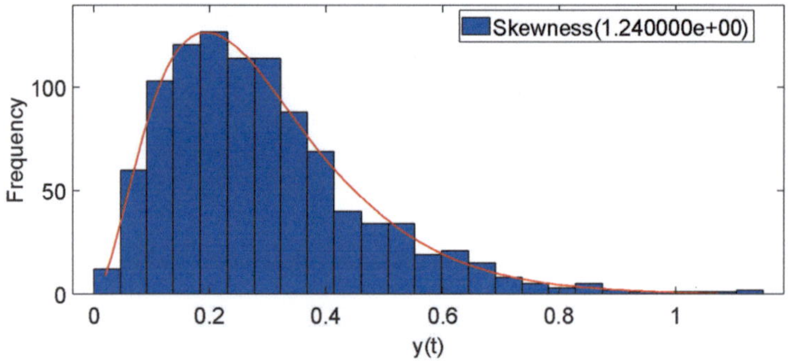

Fig. 1.20 Histogram of a time series with positive skewness

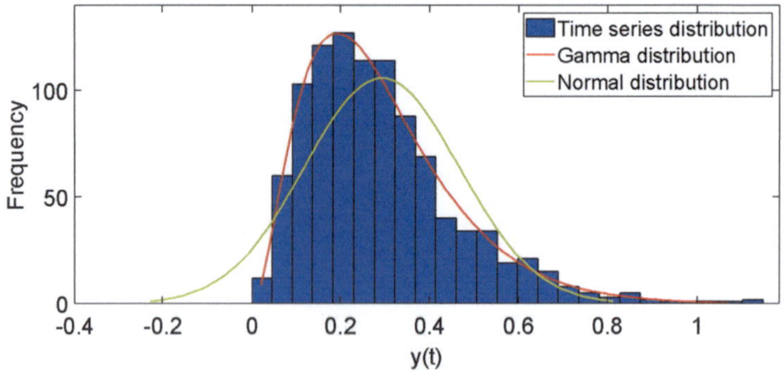

Fig. 1.21 Histogram of a time series with positive skewness, marked with the gamma and normal distribution

Weibull Distribution

In the previous paragraph, we have seen an example of a time series with positive skewness. Now we are going to see an example of a time series where the distribution of the data presents a negative skewness which is a characteristic of the well-known Weibull probability distribution which is described by the following equation:

$$f(x)=\frac{\gamma}{\alpha}\left(\frac{x-\mu}{\alpha}\right)^{(\gamma-1)}\exp\left(-\left(\frac{x-\mu}{\alpha}\right)^{\gamma}\right)x\geq\mu;\gamma,\alpha>0 \tag{1.11}$$

```
%Script 1.13.1
% Generates Weinbull random numbers

r3=wblrnd(3,45,1,1000);
plot(r3,'.-')
xlabel('Time (t)');
ylabel('y(t)');
legend('Simulated time series','Location','Best')
```

By running in the command window script 1.13.1, we plot the simulated time series as we can see in Fig. 1.22.

By running script 1.14, we obtain the statistics and the histogram of the above time series, along with a comparison with the normal distribution (Fig. 1.23, Table 1.7).

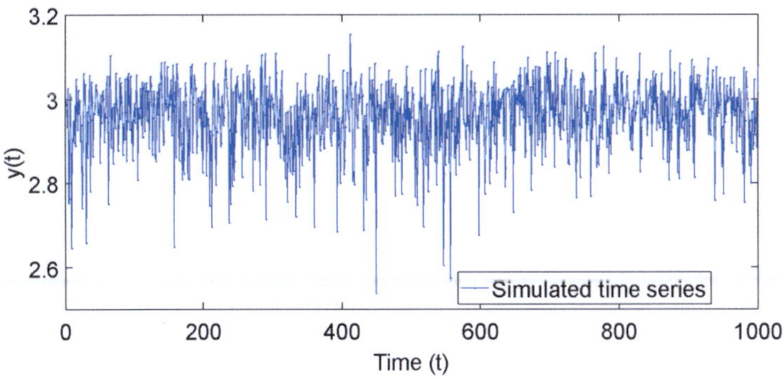

Fig. 1.22 Simulated the time series following the Weibull distribution

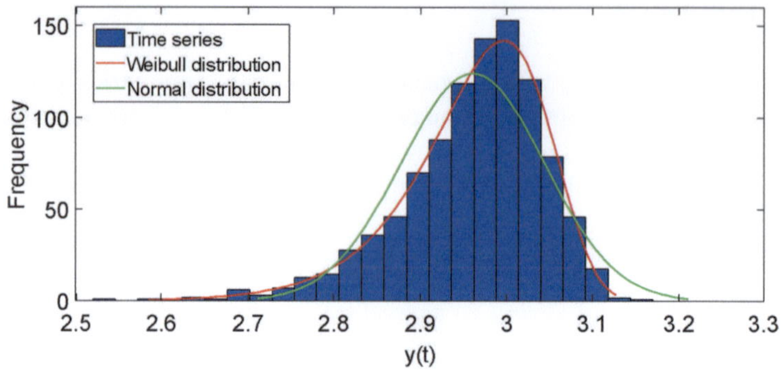

Fig. 1.23 Histogram of a time series with negative skewness, marked with the Weibull and normal distribution

Table 1.7 Descriptive statistics

Mean	Std	Var	Skew	Kurtosis	Median	Max	Min
2.960	0.083	0.007	−1.027	4.854	2.974	3.154	2.539

```
% Script 1.14
% Histogram of time series with negative skewness

TS=input('Give the time series_:'); % i.e r3
Skewness = input('Give the skewness_:'); % i.e -1.027 skewness of
r3
h = histfit(TS',25,'weibull');set(h(1),'color','b');
set(h(2),'color','r')
ylabel('Frequency');
leg1=legend(sprintf('Time series Skewness(%d)',Skewness),'Weibull
distribution','Location','Best')
hold on
h1=histfit(TS',25,'normal');set(h1(1),'color','b');
set(h1(2),'color','g')
```

Although statistical analysis is very useful and provides a certain insight for the system under study, there are some aspects that cannot be resolved such as, for example, any temporal relation or the existence of deterministic laws that could help describe the phenomenon and perhaps permit prediction of its behavior.

1.4 Components of a Time Series

This subsection presents the basic information about the components of a time series and the methods to decompose it. Time series are affected by the following three components:

- **Trend** refers to the tendency of the time series to increase, decrease, or remain constant over a long period of time. Otherwise, it indicates the long-term change in the mean level of data. The trend may be linear or nonlinear and may vary over time.
- **Seasonality/periodicity** shows a repeating pattern present in a time series. Several factors can cause seasonal variations, such as seasonal effects in environmental data, or certain times, such as the end of the year for financial data, etc.
- The third component is the **irregular behavio**r which is combined with the previous components. This component may be random/stochastic, or it may follow a more complex behavior as it is the case in chaotic systems.

Seasonality and trend are considered as deterministic components.

Two different types of models are generally used to describe the time series considering the effect of these four components.

The **additive model** is the model that time series is supposed to be the sum of the three components:

$$X(t) = T(t) + S(t) + I(t) \tag{1.12}$$

where $T(t)$, $S(t)$, and $I(t)$ are trend, seasonal, and irregular components, respectively, at time (t).

The **multiplicative model** is the model where the time series is the product of the three above components:

$$X(t) = T(t)S(t)I(t) \tag{1.13}$$

In order to demonstrate the above concepts, we present in Fig. 1.24 a signal that is synthetic and originates from three components: trend, periodicity, and irregular component, produced using script 1.15.

```
%%
%Script 1.15
%Generate time series with noise and trend

N=input('Give the time series length_:');300
f=input('Give the frequency_:');  10,10,0.2
Amp=input('Give the Amplitude_:');
t=0:1:N;
y=Amp*sin(2*pi*f*t/500);

noise=input('Give the noise level_:');
ynoise=AddNoise(y',noise);  % add noise

t=1:N+1;
Ynoise_trend=(ynoise' + t/10)';
plot(Ynoise_trend,'b.-','MarkerSize',10);
ylim([-Amp-5 4*Amp])
xlim([0 N+1])
xlabel('Time (t)');
ylabel('y(t)');
legend('Time series with noise and trend','Location','Best')
```

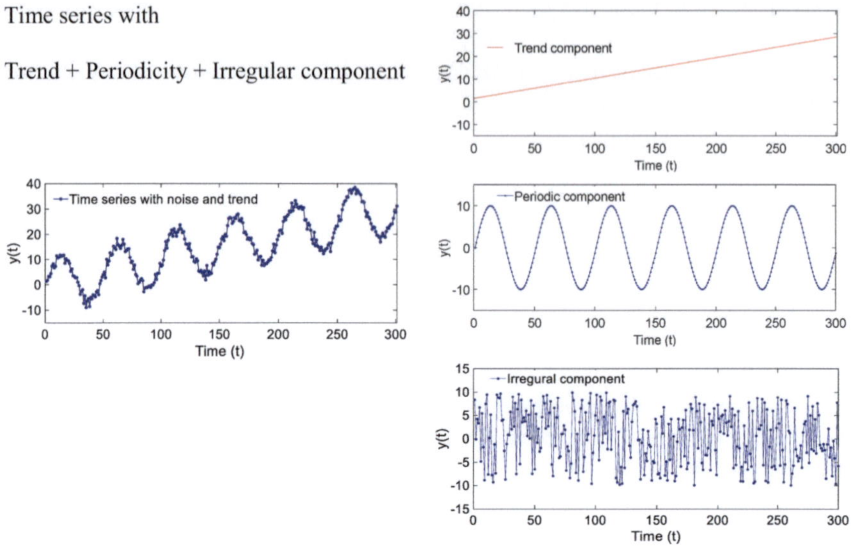

Fig. 1.24 A time series with noise, periodicity, trend (left) and the time series of the three components (on the right)

By employing Matlab scripts 1.15 and 1.16, we get the time series of Fig. 1.24.

```
%%
%Script 1.16
%Generate periodic time series

N=input('Give the time series length_:');
f=input('Give the frequency_:');
Amp=input('Give the Amplitude_:');
t=0:1:N;
y=Amp*sin(2*pi*f*t/500);
plot(y,'b.-','MarkerSize',10);    %plot function
ylim([-Amp-5 Amp+5])
xlim([0 N])
xlabel('Time (t)');
ylabel('y(t)');
legend('Periodic component ','Location','Best')
```

The main purpose of the analysis of a time series, through the process of decomposition into individual time series which when synthesized create the time series, is the study of the irregular component. This component may contain important information for explaining the time series under study.

Another example of a time series with trend is presented in Fig. 1.25 and is obtained using script 1.17.

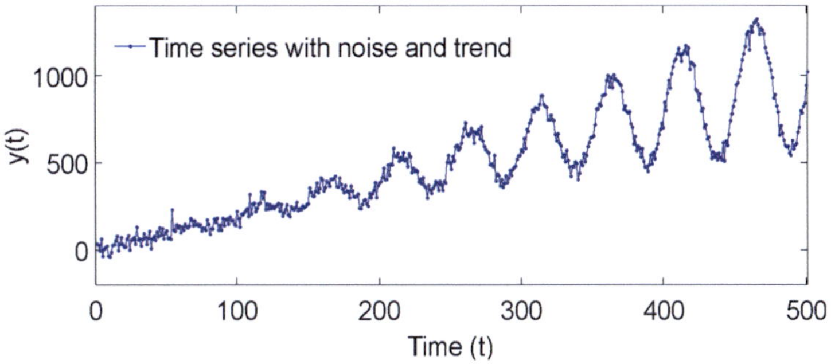

Fig. 1.25 Time series with noise, periodicity, and trend

```
%%
%Script 1.17
%Generate time series with noise and trend

N=input('Give the time series length_:');
t=0:1:N;
A=sqrt(t)/3;
y=A.*A.*A.*sin(2*pi*t/50);

noise=input('Give the noise level_:');
ynoise=AddNoise(y',noise);

t=1:N+1;
Ynoise_trend=(ynoise' + 2*t)';  % TStrend the new name of time
series
plot(Ynoise_trend,'b.-','MarkerSize',10);   %plot function
xlim([0 N+1])
xlabel('Time (t)');
ylabel('y(t)');
legend('Time series with noise and trend','Location','Best')
```

```
%%
%Script 1.18
%Generate time series

N=input('Give the time series length_:');
t=0:1:N;
A=sqrt(t)/3;
y=A.*A.*A.*sin(2*pi*t/50);
plot(y','b.-','MarkerSize',10);   %plot function
ylim([min(y)-5 max(y)+5])
xlim([0 N+1])
xlabel('Time (t)');
ylabel('y(t)');
legend('Time series','Location','Best')
```

Using script 1.18 and giving the length of the time series, we get the time series of Fig. 1.26.

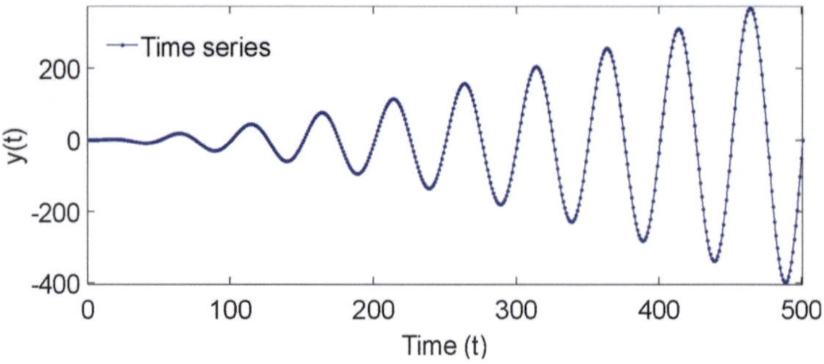

Fig. 1.26 Periodicity component

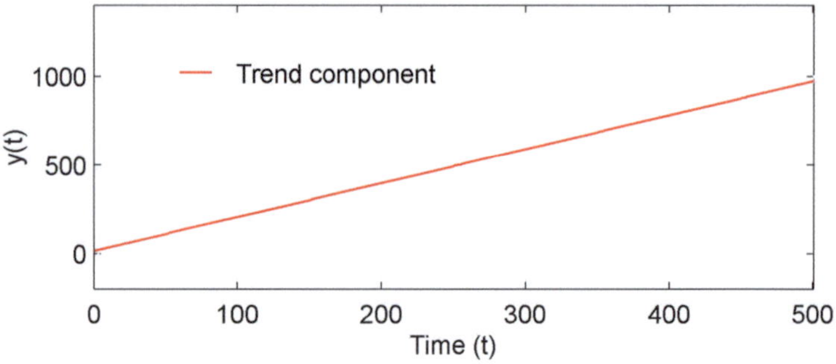

Fig. 1.27 Trend component

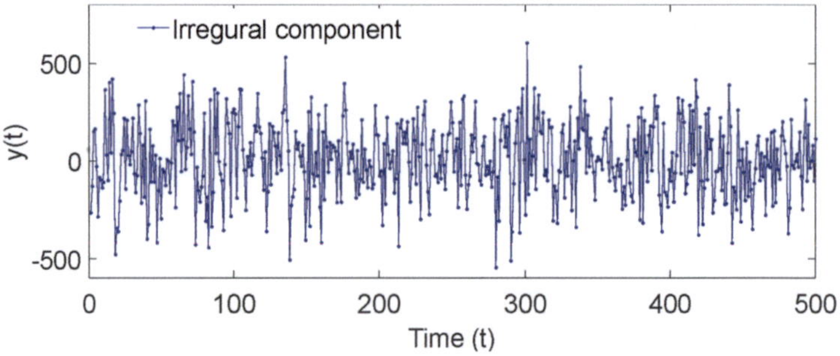

Fig. 1.28 Irregular component

Figure 1.27 presents the trend component, and Fig. 1.28 the irregular component which is added to the time series according to script 1.17.

The question that naturally arises is how we can decompose and detect these components in a given time series. There is also another reason that we need to detect these components/behaviors. When we have trend and/or seasonality in a time series, it means that the statistical measures that we extract (mean, standard deviation, etc.) are not constant all along the time of observation and thus the time series is not stationary. However, many tests and methods suppose that the time series analyzed is **stationary**. We say that a time series presents **strong stationarity** when the descriptive statistical measures are constant along the time series (i.e., mean, variance, kurtosis, and skewness) while we say that it presents **weak stationarity** when only the mean and variance are constant along the time series. The latter is the most common case.

1.4.1 Trend/Seasonal Component (Period Estimation)

The first component, trend, shows the tendency of the variables to increase or decrease as the data evolves over time. Uptrends represent the increase of the data over time, and downtrends show the decrease.

One of the simplest tests to determine if a time series presents a constant trend is to divide it into smaller segments and calculate the change in the mean value of the individual segments. If the average values are not constant in the evolution of the time series, then we can say that the time series has a trend.

We can see the time series of Fig. 1.29.

Using script 1.19, we obtain the mean in successive segments, as shown in Fig. 1.29. We can clearly see that the time series presents a trend since the average value of the data changes as a function of the interval, and more specifically, it seems to increase at a rather constant rate (Fig. 1.30).

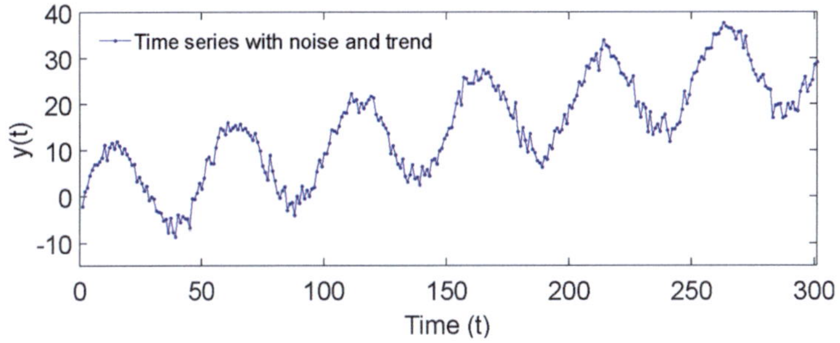

Fig. 1.29 Time series with noise and trend

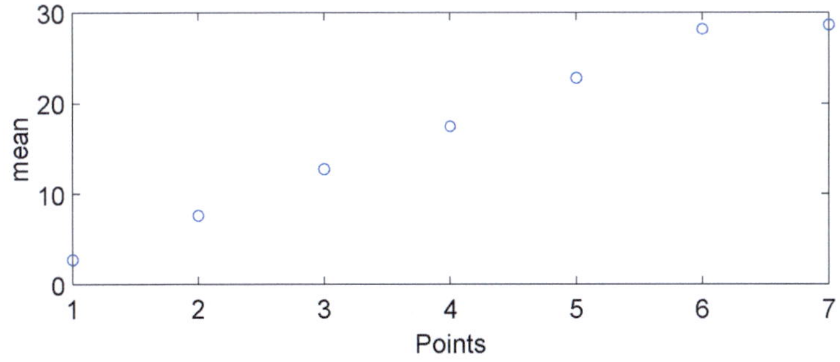

Fig. 1.30 Successive segments mean of the time series of Fig. 1.29

```
%Script 1.19
% Test for trend using the mean value

TS=input('Give the time series_:');  %Ynoise_trend
Size_segment=input('Give the time series (segment) length_:');
%50
Overlap=input('Give the overlap of segments_');  %0

[TS_segments,index,reject] =slideWindow(TS, Size_segment,
Overlap);
TS_segments(TS_segments==0)=NaN;
columnMeans = mean(TS_segments,'omitnan');
plot(columnMeans,'o')
xlabel('Points');
ylabel('mean');
```

Detrending is the process of removing the trend from a data set. It is a crucial step of time series analysis. There are several detrending methods described below.

1.4.2 *Detrending and De-Seasoning of a Time Series*

Detrending and De-seasoning are techniques to remove trend and seasonal components from a time series.

Detrending Using a Fit Deduced Model

To be able to understand the trend in time series, we will use the examples mentioned earlier. As we have seen, there is rather a linear trend so we can try to fit a linear model of the form $x = at + b$, subtract it from the time series and study the remaining part.

The following script 1.20 shows an example of a linear fit application. When subtracted from the time series, we have the remaining oscillating part with a large period indicating seasonality along with small variations on top of it. Using script 1.20, the initial and the detrended time series are presented in Fig. 1.31.

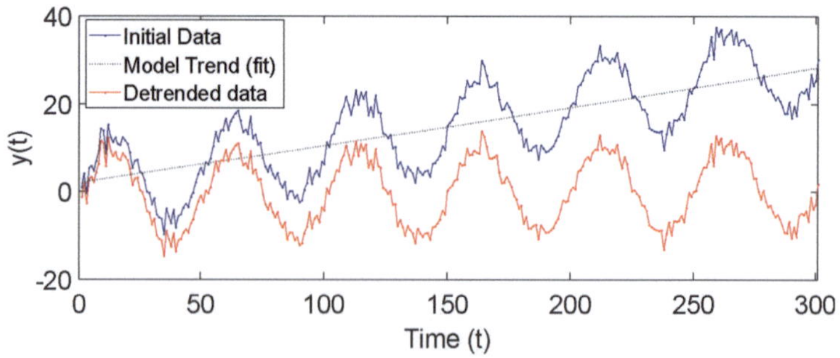

Fig. 1.31 Initial data, model trend, and detrended data

```
%% Script 1.20
% Detrend data with fit model

TStrend = input('Give the time series_:');
length = input('Give the time series lenght_:');
Degreefit=input('Give the degree polynomial_:');
t=(1:length)';
p=polyfit(t,TStrend,Degreefit);
f=polyval(p,t);
plot(t,TStrend,'b.-',t,f,':k')
hold on
TSdetrend=TStrend-f;
plot(TSdetrend,'r.-');
axis([0 length -20 40])
legend('Initial Data','Model Trend (fit)','Detrended data')
xlabel('Time (t)');
ylabel('y(t)');
```

Detrending Using Moving Average Mean

Another method to subtract the time series trend is to apply the rolling average function, detrending moving average (DMA) to the time series data and then subtract the average result from the time series. The DMA consists of variance $\sigma^2_{DMA}(n)$ of the time series $y(i)$ $i = 1$ to N with the respect to the trend $y_n(i)$ at scale n:

$$\sigma^2_{\mathrm{DMA}}(n) = \frac{1}{N-n+1}\sum_i \left[y(i) - \tilde{y}n(i)\right]^2 \tag{1.14}$$

where $\tilde{y}n(i)$ is defined as a time dependent average function of $y(i)$ and

$$\tilde{y}_n(i) = \frac{1}{n}\sum_{k=0}^{n-1} y(i-k) \tag{1.15}$$

In this case, what needs to be taken care of is not to remove from the time series a component/element that contains useful information. To prevent this from happening, it is advisable to use as many averaging points in the function as the time series period.

In the following, we present some examples.

```
%Script 1.21
% Detrend data with applying moving average (1) TSTN

TStrend = input('Give the time series with trend_:');
ma_mean = input('Give the moving mean k_:');
m = movmean(TStrend,ma_mean);
TSdetrend = TStrend - m;
figure
plot(TStrend,'.-')
hold on
plot(m,'-r')
plot(TSdetrend,'.-')
xlabel('Time (t)');
ylabel('y(t)');
legend('Time series with trend',sprintf('Moving Average Mean(%d)
degree',ma_mean),'Detrended','Location','Best')
title(sprintf('Detrended time series by Moving Average(%d)
mean',ma_mean))
```

Using script 1.21 in this example, we chose 30 points for a moving average as the time series has a periodicity of about 30 points and the results appear in Fig. 1.32.

In Fig. 1.33, we present the results if we choose 10 points for averaging (using script 1.21). As we can see this choice results in a smoother curve. From the figure, we observe that not only is the trend removed, but also useful information about the time series.

Detrending Using Moving Average Model (Filter)

The following code (script 1.22) shows how to remove a trend component from a time series using moving average mean model (filter).

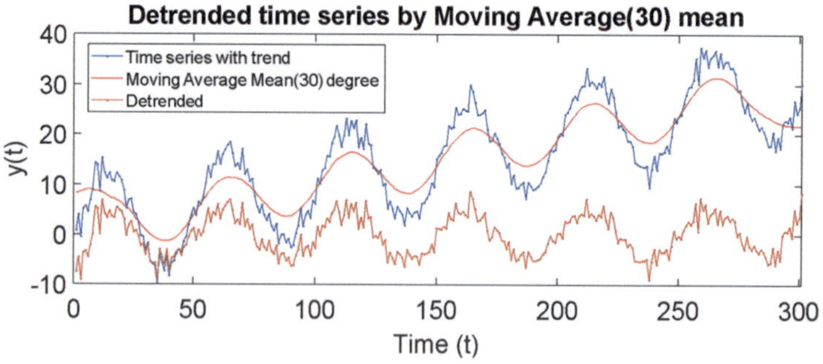

Fig. 1.32 Detrended time series using moving mean of 30 points

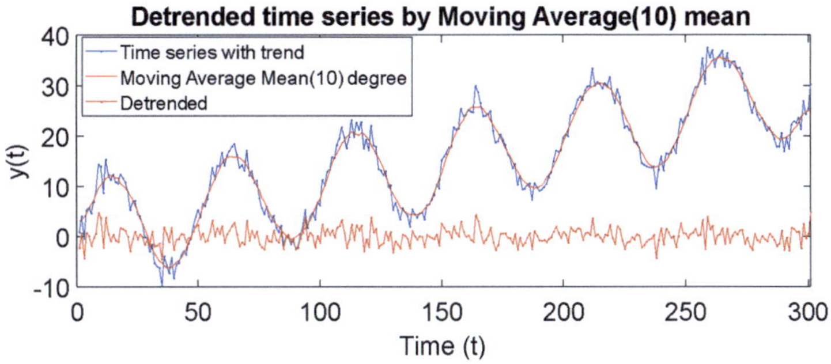

Fig. 1.33 Detrended time series using moving mean of 10 points

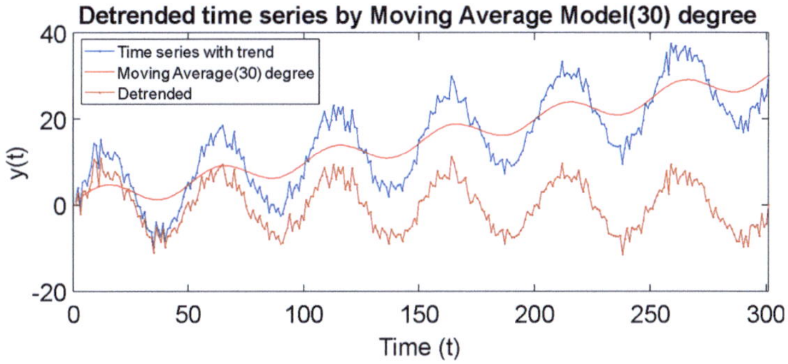

Fig. 1.34 Detrended time series using moving model (filter) of 30 points

```
% Script 1.22
% Detrend data with applying moving average model TSTN

TStrend = input('Give the time series with trend_:');
ma_degree = input('Give the moving average degree_:');
m = ones(1,ma_degree)/ma_degree;
mafit = filtfilt(m,1,TStrend);
figure
plot(TStrend,'.-')
hold on
plot(mafit,'-r')
xlabel('Time (t)');
ylabel('y(t)');
TSdetrend = TStrend - mafit;
hold on
plot(TSdetrend,'.-')
xlabel('Time (t)');
ylabel('y(t)');
legend('Time series with trend',sprintf('Moving Average(%d)
degree',ma_degree),'Detrended','Location','Best')
title(sprintf('Detrended time series by Moving Average Model(%d)
degree',ma_degree))
```

By running script 1.22, we obtain the results presented in Fig. 1.34.

If we change the number of moving average points from 30 to 10, we get the results appearing in Fig. 1.35. As we can see this choice results in a smoother curve. From the figure, we observe that not only is the trend removed, but also useful information about the time series.

Detrending Using First Differences

A common method to remove trends is to transform the time series employing first differences:

$$y(t)' = y(t) - y(t-1) \tag{1.16}$$

where $y(t)$ $t = 1$ to N is the original time series.

This procedure can be applied using script 1.23, and the results of using it appear in Fig. 1.36.

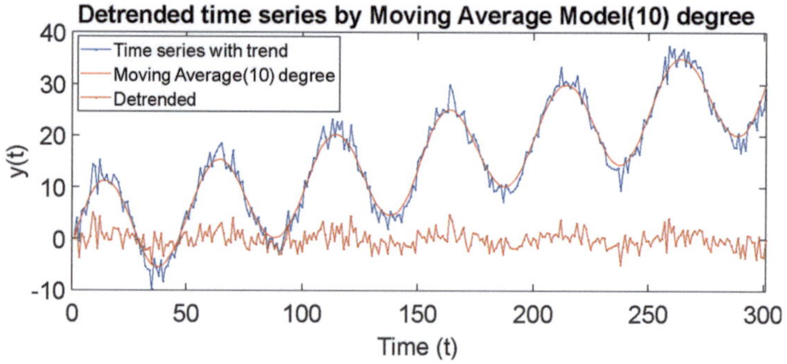

Fig. 1.35 Detrended time series using moving model (filter) of 10 points

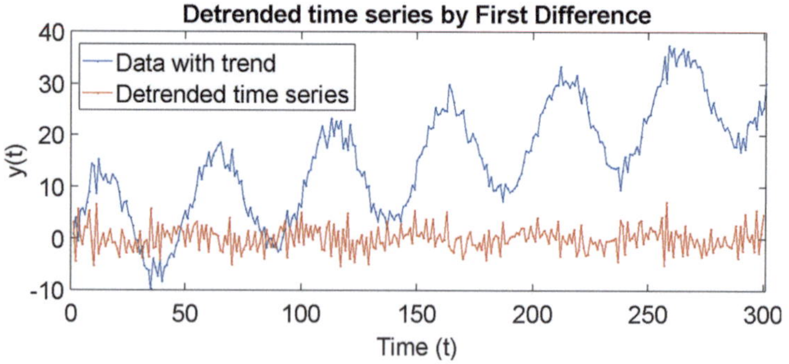

Fig. 1.36 Original and detrended time series using first differences

```
%%Script 1.23
% Detrend data with applying first difference TSTN

TStrend = input('Give the time series with trend_:');
length = input('Give the time series length_:');
t=1:length;
Diff_detreded=diff(TStrend)    % Diff_y1 the time series without
trend
plot(t,TStrend,'.-');
hold on
plot(Diff_detreded,'.-');
legend('Data with trend','Detrended time series')
xlabel('Time (t)');
ylabel('y(t)');
title('Detrended time series by First Difference','FontSize',14)
```

In the case of the first differences, the trend is removed, and we see that to some extent, the periodicity of the time series is maintained.

In the following, we present examples of time series with nonlinear trend as well as application of trend removal methods.

Example with Nonlinear Trend

In Fig. 1.37, we can see an example of time series presenting a nonlinear trend. This is a synthetic time series produced following the next steps.

First, using script 1.23.1, we add a nonlinear component to a time series. We obtain the results appearing in Fig. 1.37.

```
%%Script 1.23.1
% Add non linear trend to time series   r

TS = input('Give the time series_:');
length = input('Give the time series lenght_:');
t=1:length;
nonlinear_trend = (0.01 * t.^2 - 0.5 * t + 2)';
TStrendnon=[TS + nonlinear_trend];
figure
plot(t,TS,t,TStrendnon)
legend('Initial Data','Data with nonlinear Trend')
xlabel('Time (t)');
ylabel('y(t)');
```

Now, we have a new time series (the red one) with a nonlinear trend. We can try to remove the nonlinear trend by applying the first differences technique (script 1.23).

The results are depicted in Fig. 1.38.

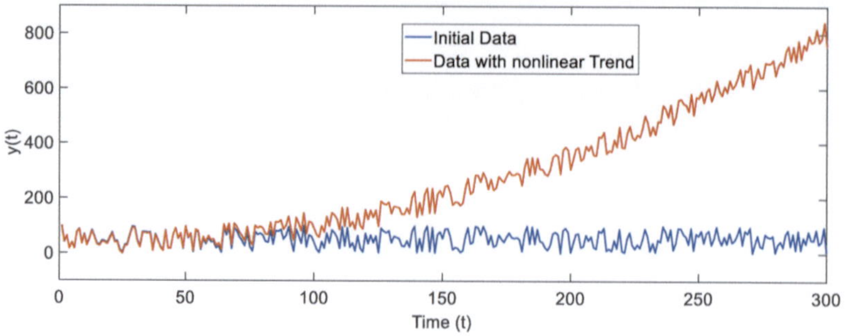

Fig. 1.37 Original and detrended time series using first difference

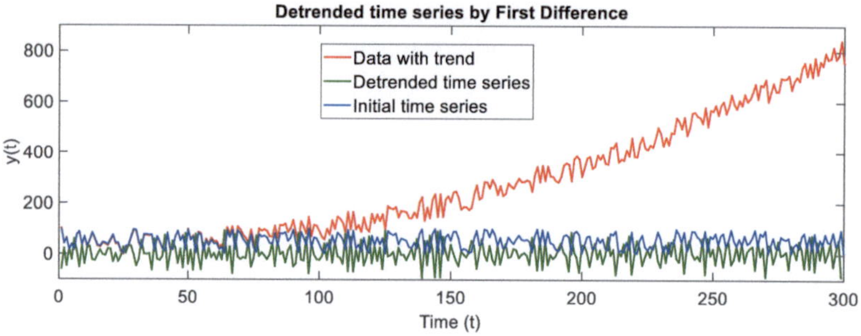

Fig. 1.38 Original and detrended time series using first difference

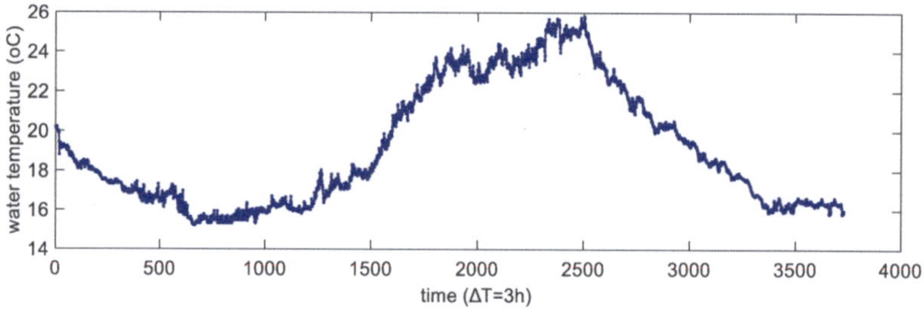

Fig. 1.39 Sea water temperature time series (Charakopoulos et al. 2018)

1.4.3 Detrending and De-Seasoning of a Real-Time Series

The above methods are then applied to real-world time series. Figure 1.39 shows the sea water temperature time series [1].

We first applied script 1.19 to evaluate whether the data exhibit a trend. The results are presented in Fig. 1.40, which shows that the values of the successive means differ. This indicates that the time series displays a trending behavior and is not stationary.

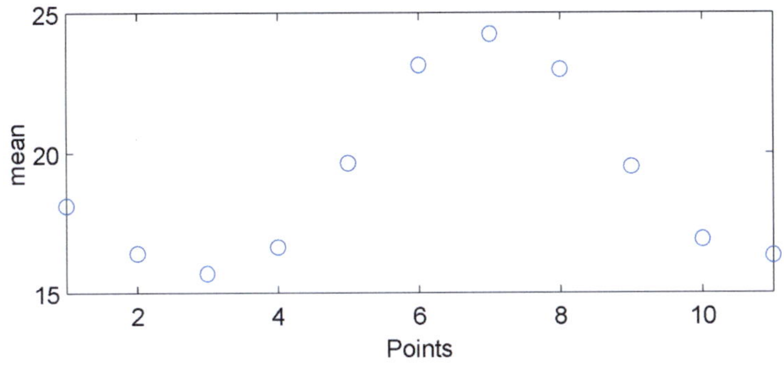

Fig. 1.40 Successive mean of the time series of Fig. 1.39

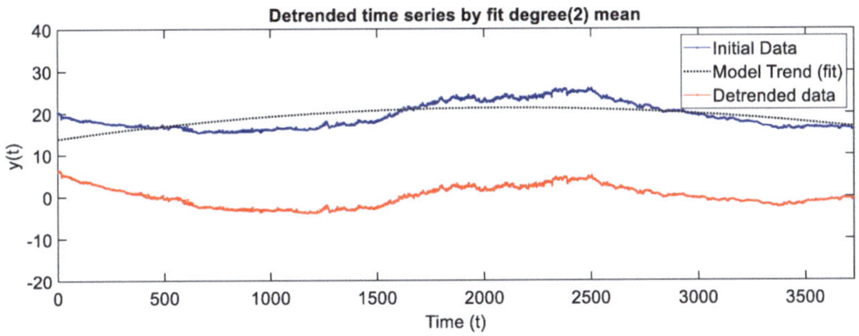

Fig. 1.41 Original and detrended time series using second-degree polynomial fit

To remove the trend from the data, we first applied the polynomial fit method, as the trend is nonlinear. In this context, it is crucial to carefully select the appropriate degree of the polynomial model using script 1.20. The results are shown in Fig. 1.41, where it is evident that the trend remains in the data because the correct polynomial degree was not chosen.

We can notice that the choice of the second-degree polynomial fit is not suitable for the time series. Therefore, a higher degree of polynomial should be used.

In Fig. 1.42 shows the results using an eighth-degree polynomial fit. We can observe that by choosing a different degree model, we can extract the trend of the time series.

Then, for the same time series, the moving average method is applied. Using the following script 1.24, we obtain the results presented in Figs. 1.43 and 1.44.

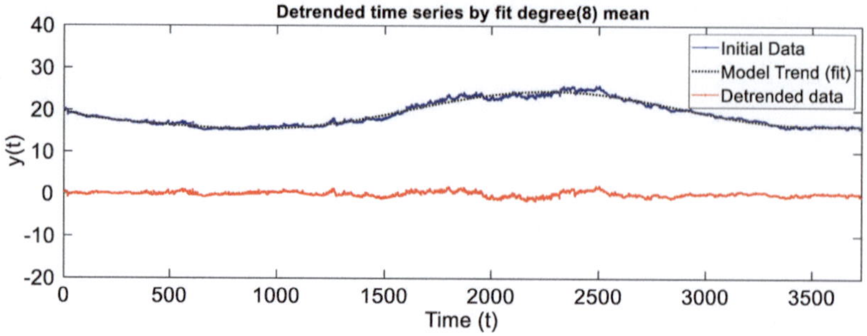

Fig. 1.42 Original and detrended time series using eighth-degree polynomial fit

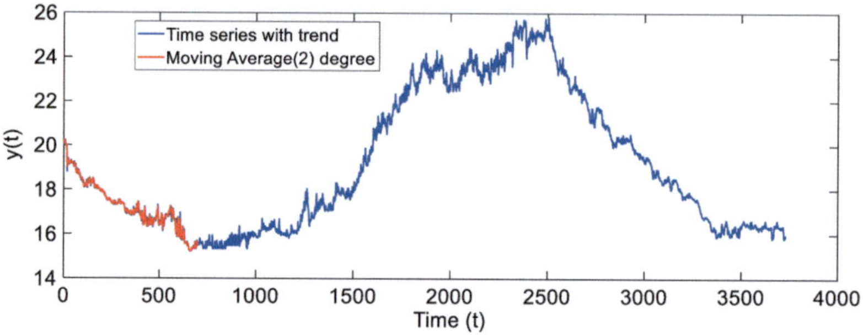

Fig. 1.43 Sea water temperature time series and the applied second degree model

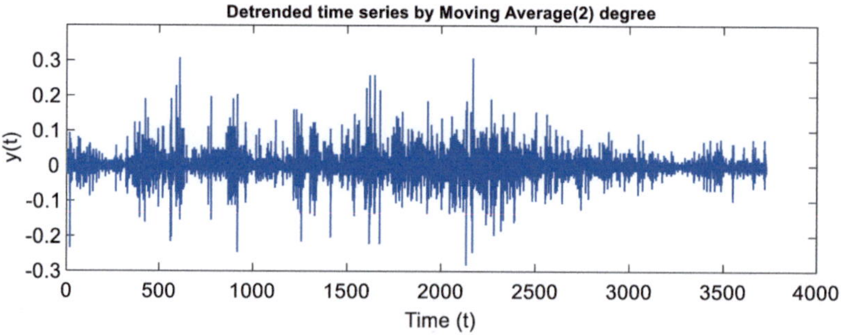

Fig. 1.44 Detrended sea water temperature time series

```
% script 1.24
%% Detrend data with applying moving average

TStrend = input('Give the time series with trend_:');
ma_degree = input('Give the moving average degree_:');
m = ones(1,ma_degree)/ma_degree;
mafit = filtfilt(m,1,TStrend);

figure
plot(TStrend,'-')
hold on
plot(mafit(1:700),'.-r')
xlabel('Time (t)');
ylabel('y(t)');
legend('Time series with trend',sprintf('Moving Average(%d)
degree',ma_degree),'Location','Best')

TSdetrend = TStrend - mafit;
figure(3)
clf
plot(TSdetrend,'.-')
xlabel('Time (t)');
ylabel('y(t)');
title(sprintf('Detrended time series by Moving Average(%d)
degree',ma_degree))
```

Figure 1.45 shows the result of applying the method (script 1.25) of the first differences in the same time series.

```
% script 1.25


%% Detrend data with applying first difference

TStrend = input('Give the time series with trend_:');
length = input('Give the time series lenght :');
t=1:length;
Diff_detreded=diff(TStrend);    % Diff_y1 the time series without
trend
plot(Diff_detreded);
legend('Detrended time series')
xlabel('Time (t)');
ylabel('y(t)');
```

We present another example of a detrended time series using the corresponding script 1.26 and **gold price time series** data. Figure 1.46 presents the results of checking whether the gold price time series contains a trend.

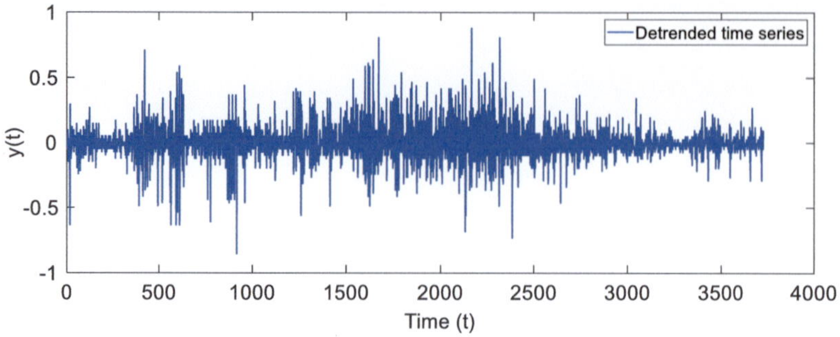

Fig. 1.45 Detrended water temperature time series using first differences

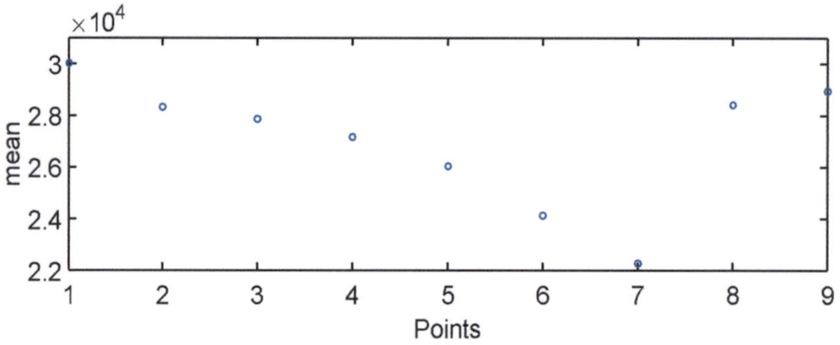

Fig. 1.46 Results of mean in sliding segments of gold time series

```
% script 1.26
%% Test for trend using the mean value

TS=input('Give the time series_:');
Size_segment=input('Give the time series (segment) length_:');
Overlap=input('Give the overlap of segments_');

[TS_segments,index,reject] =slideWindow(TS, Size_segment,
Overlap);
TS_segments(TS_segments==0)=NaN;
columnMeans = mean(TS_segments,'omitnan');
plot(columnMeans,'o')
xlabel('Points');
ylabel('mean');
```

First, we applied the detrend using fit model using the following code (script 1.27), and the results appear in Fig. 1.47.

```
% script 1.27
%% Detrend data with fit model (DJI_index)

TStrend = input('Give the time series_:');
length = input('Give the time series lenght_:');
Degreefit=input('Give the degree polynomial_:');
t=(1:length)';
p=polyfit(t,TStrend,Degreefit);
f=polyval(p,t);
plot(t,TStrend,t,f,':k')
hold on
TSdetrend=TStrend-f;          % y1detrend time series name
plot(TSdetrend);
legend('Data with Trend',sprintf('(%d) Degree
model',Degreefit),'Detrended data')
xlabel('Time (t)');
ylabel('y(t)');
```

Next, we applied the detrend method (script 1.28) using the moving average approach as it appears in the following code (script 1.28), and the results appear in Figs. 1.48 and 1.49.

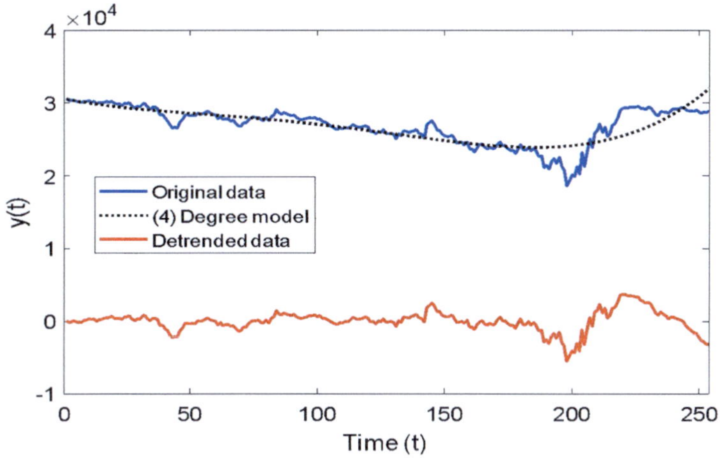

Fig. 1.47 Results of detrend using detrend fit model

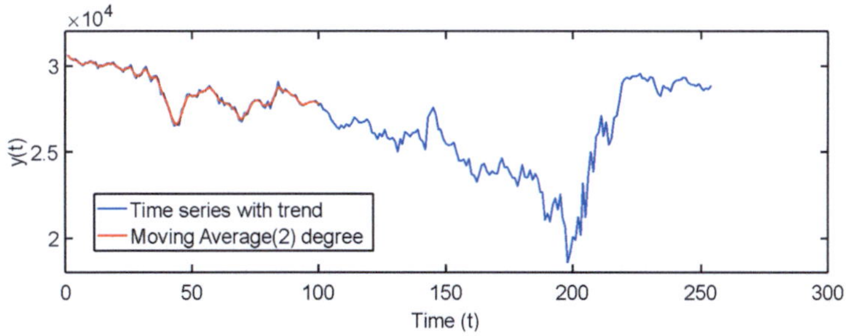

Fig. 1.48 Gold time series and fitting a moving average degree model

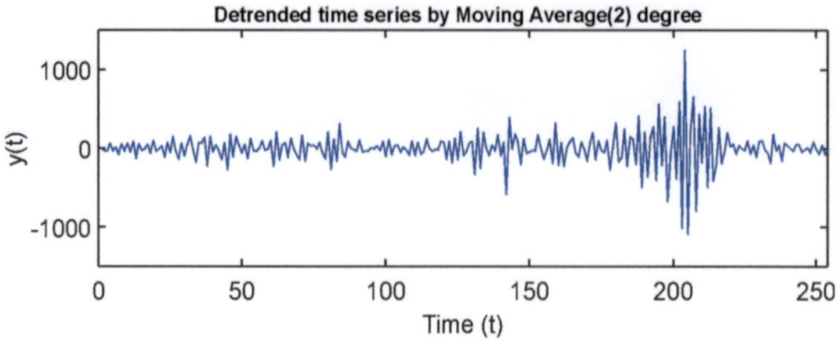

Fig. 1.49 Results of detrend using moving average degree model

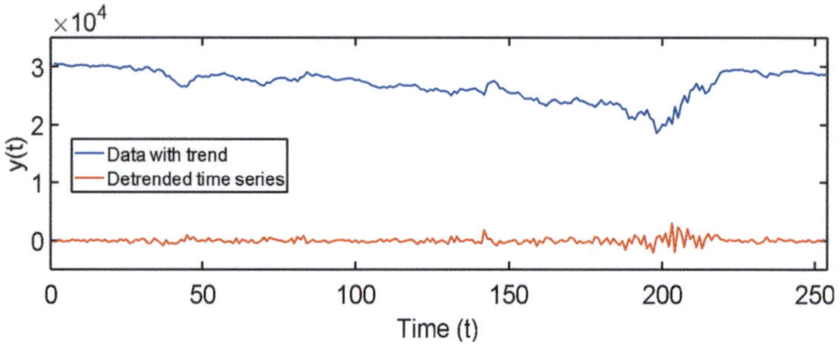

Fig. 1.50 Results of detrend using first difference method

```
% script 1.28
%% Detrend data with applying moving average (DJI_index)

TStrend = input('Give the time series with trend_:');
ma_degree = input('Give the moving average degree_:');
m = ones(1,ma_degree)/ma_degree;
mafit = filtfilt(m,1,TStrend);

figure
plot(TStrend,'-')
hold on
plot(mafit(1:100),'.-r')
xlabel('Time (t)');
ylabel('y(t)');
legend('Time series with trend',sprintf('Moving Average(%d)
degree',ma_degree),'Location','Best')

TSdetrend = TStrend - mafit;
figure(3)
clf
plot(TSdetrend,'.-')
xlabel('Time (t)');
ylabel('y(t)');
title(sprintf('Detrended time series by Moving Average(%d)
degree',ma_degree))
```

We applied script 1.29, and the results appear in Fig. 1.50.

```
% script 1.29
%% Detrend data with applying first difference

TStrend = input('Give the time series with trend_:');
length = input('Give the time series lenght_:');
t=1:length;
Diff_detreded=diff(TStrend);    % Diff_y1 the time series without
trend
plot(t,TStrend);
hold on
plot(Diff_detreded);
legend('Data with trend','Detrended time series')
xlabel('Time (t)');
ylabel('y(t)');
```

References

1. Charakopoulos, A. K., Karakasidis, T. E., & Liakopoulos, A. (2015). Spatiotemporal analysis of seawatch buoy meteorological observations. *Environmental Processes, 2*, 23–39.
2. Charakopoulos, A. K., Karakasidis, T. E., Papanicolaou, P. N., & Liakopoulos, A. (2014). The application of complex network time series analysis in turbulent heated jets. *Chaos: An Interdisciplinary Journal of Nonlinear Science, 24*(2).
3. Chatfield, C. (2013). *The analysis of time series: Theory and practice*. Springer.
4. Cryer, J. D. (2008). *Time series analysis*. Springer.
5. Hamilton, J. D. (2020). *Time series analysis*. Princeton University Press.
6. Kantz, H., & Schreiber, T. (2003). *Nonlinear time series analysis*. Cambridge University Press.
7. Karakasidis, T. E., & Charakopoulos, A. (2009). Detection of low-dimensional chaos in wind time series. *Chaos, Solitons & Fractals, 41*(4), 1723–1732.
8. Kirchgässner, G., Wolters, J., & Hassler, U. (2012). *Introduction to modern time series analysis*. Springer Science & Business Media.
9. Pham, H. (Ed.). (2023). *Springer handbook of engineering statistics*. Springer Nature.
10. Shumway, R. H., Stoffer, D. S., & Stoffer, D. S. (2000). *Time series analysis and its applications* (Vol. 3, p. 4). Springer.

Chapter 2
Temporal Behavior of Time Series

Time series analysis focuses on understanding the temporal behavior of data points collected sequentially over time. This involves exploring patterns such as trends, seasonality, and cyclicity, as well as deeper properties like autocorrelation, which quantifies the similarity between observations as a function of time lag, and the power spectrum, which reveals the frequency components of the data. Advanced techniques, such as calculating mutual information, help uncover nonlinear dependencies, while measures like the Hurst exponent provide insights into long-term memory and persistence within the series. Additionally, Hjorth parameters offer a comprehensive framework for characterizing the activity, mobility, and complexity of time-varying signals. By integrating these concepts, researchers can gain a multidimensional understanding of time series, enabling better predictions, enhanced feature extraction, and deeper exploration of the underlying dynamics [3, 4].

2.1 Autocorrelation

Autocorrelation measures the degree of similarity between a time series and its lagged version over consecutive time intervals. It quantifies the relationship between a variable's present value and its past values.

The autocorrelation function (ACF) assesses the correlation between observations in a time series for a set of lags. The ACF for time series y is given by:

$$r_k = \frac{\sum_{t=k+1}^{n-k}\left(x_{t-k} - \overline{x}\right)\left(x_t - \overline{x}\right)}{\sum_{t=1}^{n}\left(x_t - \overline{x}\right)^2} \quad k = 1, 2, \ldots \tag{2.1}$$

T. Karakasidis, A. Charakopoulos, *Time Series and Networks Analysis*,
https://doi.org/10.1007/978-3-031-92628-0_2

Here, r_k represents the ACF correlation coefficient of the series with its k lag; and n, x_t, $\underline{x}$ denote, respectively, the number of observations of the series, the t-th observation of the series, and the mean.

A positive correlation between two values signifies that when one increases, the other increases too or when one decreases, the other decreases too. Negative correlation signifies that when a value increases/decreases, the other decreases/increases, respectively. The value of r_k ranges from -1 to $+1$. A value of the autocorrelation coefficient of $+1$ represents a perfect positive correlation, while a value -1 represents a perfect negative correlation. A value close to zero signifies that no correlation (or very mall correlation) is present.

Technical analysts can use autocorrelation to assess the impact of past prices of a security on its future price. However, this is not completely true since autocorrelation indicates the correlation between two observations at different points in a time series. A possible cause-result effect can be checked using other methods such as the Granger causality.

The ACF is a useful tool for identifying lags with significant correlations, helping to understand the patterns and properties of a time series. This information can then be used to model the time series data effectively. From the ACF, you can assess the randomness and stationarity of a time series, as well as identify trends and seasonal patterns.

For random data, the autocorrelations should be close to zero for all lags, a condition often referred to as white noise. In contrast, non-random data will exhibit at least one significant lag.

Using script 2.1, one can calculate the autocorrelation function and plot it in Matlab.

```
%%
%Script 2.1

data=input('Time series name_'); %r
lag_t=input('Time Lag_');
figure
subplot(2,1,1);
plot(data,'b.-','MarkerSize',6);
title('Time series','FontSize',18)
ylabel('y(t)')
xlabel('(t) Time')
subplot(2,1,2);
[acf,lags,bounds]=autocorr(data,lag_t);
autocorr(data,lag_t);
title('Sample Autocorrelation Function','FontSize',16)
legend('acf','upper bound','down bound')
ylabel('Sample autocorrelation')
xlabel('Lag')
```

For random data, autocorrelations should be near zero for all lags (Fig. 2.1).

The lines indicate the confidence intervals for non-zero values. We observe that, apart from lag zero, all values fall within these bounds, signifying insignificance.

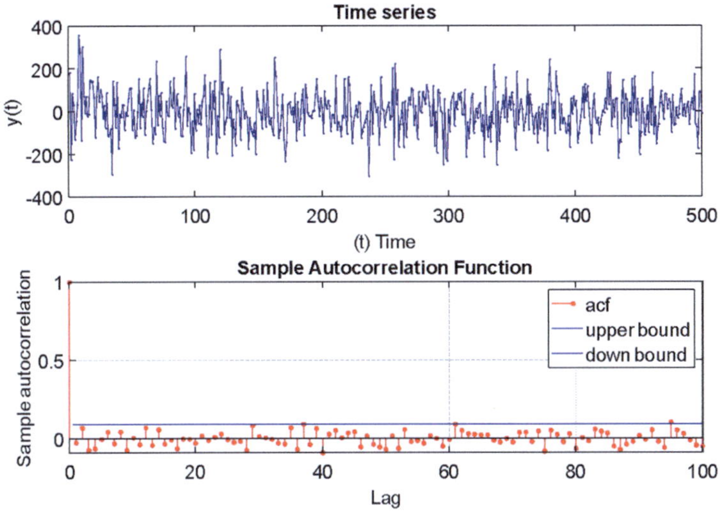

Fig. 2.1 Time series and results of autocorrelation function

We note that all prices are below the significance threshold, indicating that the values of the time series are uncorrelated and unrelated to each other, as expected for random data.

2.1.1 Seasonality Effects

A seasonal series presents alternating patterns of positive and negative autocorrelation lags. In Fig. 2.2, we present a periodic time series and its corresponding autocorrelation function. The correlation coefficients of successive values of time lags are reflected in the graph of autocorrelation function. Positive and negative autocorrelation values are observed in the periodic time series.

However, when such a behavior is observed for a long time (and we have no trends in the time series), the analyst chooses a smaller lag corresponding to autocorrelation value of 1/e (i.e., a value 0.376).

Examples of Time Series with More than One Frequency

The next example shows the application in a synthetic time series with three frequencies (Fig. 2.3). A repeating pattern is observed.

While in Fig. 2.4, we present results for three different sample rates frequencies, in order to see the effect of frequency on the autocorrelation graph.

In Fig. 2.4, we show the effect of data frequency on autocorrelation diagrams.

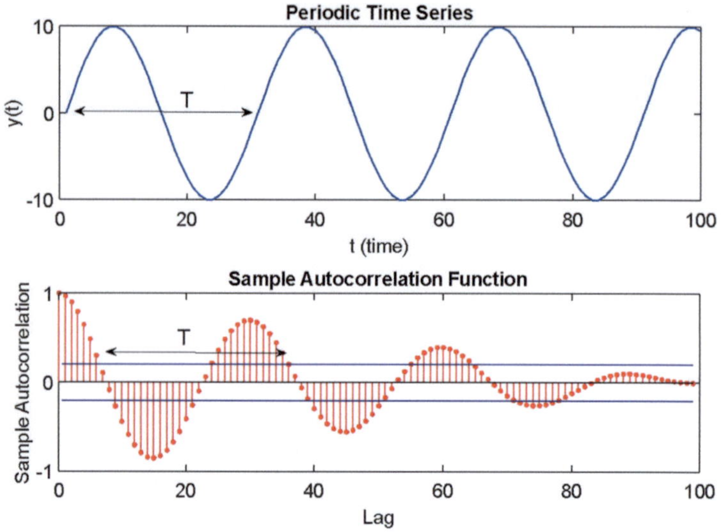

Fig. 2.2 Periodic time series and results of autocorrelation function

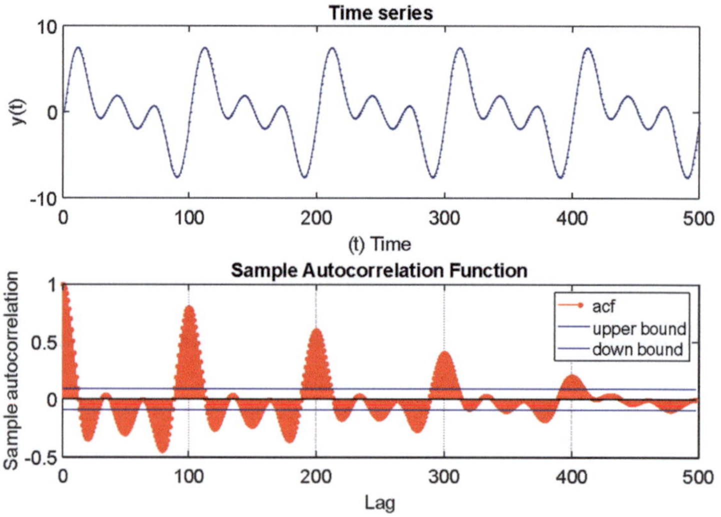

Fig. 2.3 Periodic time series with three frequencies and results of autocorrelation function

2.1.2 Noise Effects

Subsequently, it is examined whether the existence of noise affects the result of the function. In Fig. 2.5, we have results for time series without noise and time series with noise. We can see that in case the data include noise, the autocorrelation curve has lower values at the corresponding lags.

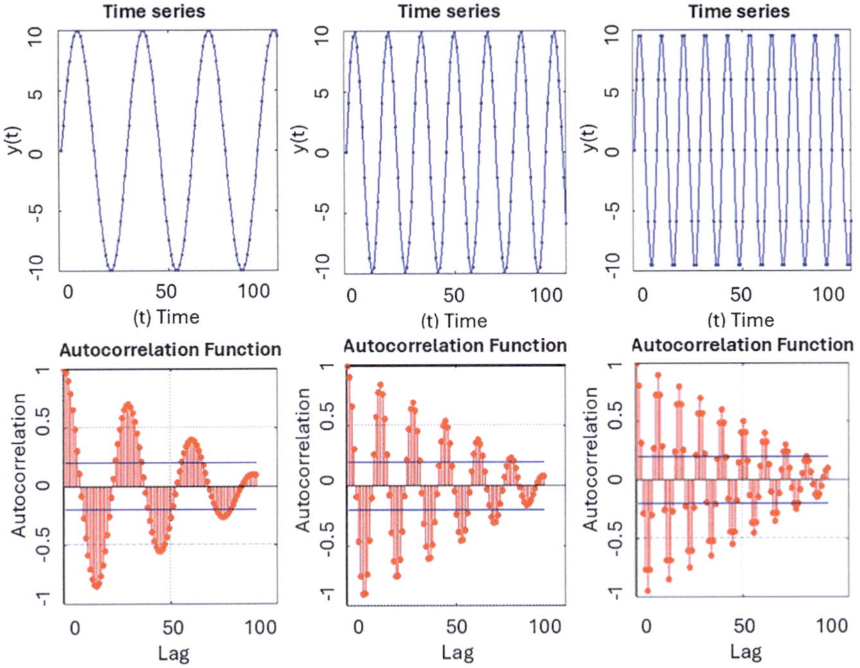

Fig. 2.4 Periodic time series with three different frequencies and the respective results of autocorrelation function

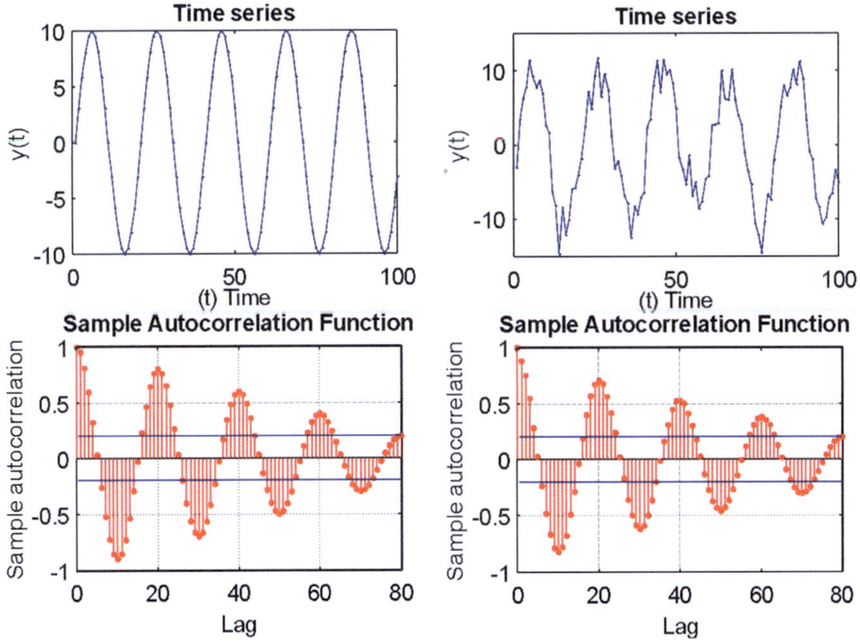

Fig. 2.5 Autocorrelation results in noisy time series

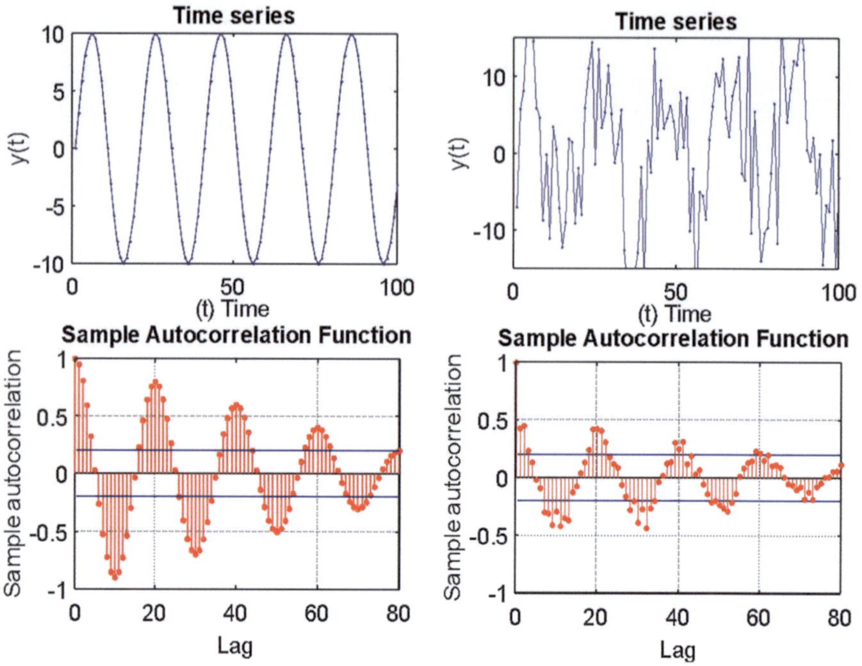

Fig. 2.6 Autocorrelation results with different level of noisy

As we can see the results are affected by the percentage of noise in the original time series. As the level of noise increases, the results of autocorrelation function are different (see Fig. 2.6). It is noted that the larger the noise rate in the initial time series, the greater the effect on the result of the autocorrelation function.

Next, we examine the effects on autocorrelation for a time series that presents trend (Fig. 2.7).

In addition, examples of calculation of the autocorrelation function in the event of deterministic time series from simulations or field measurements are presented. In Fig. 2.8, we present results for the Lorenz dynamical systems, while in Fig. 2.9, we present results for wind time series collected in the field.

As we can see without any trend removal, there are long lasting correlations.

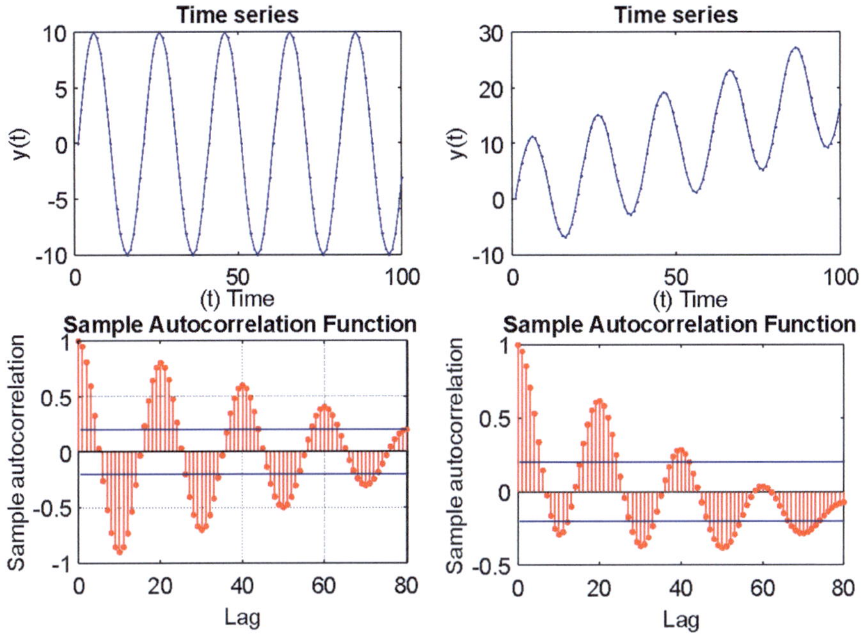

Fig. 2.7 Autocorrelation results from time series with trend

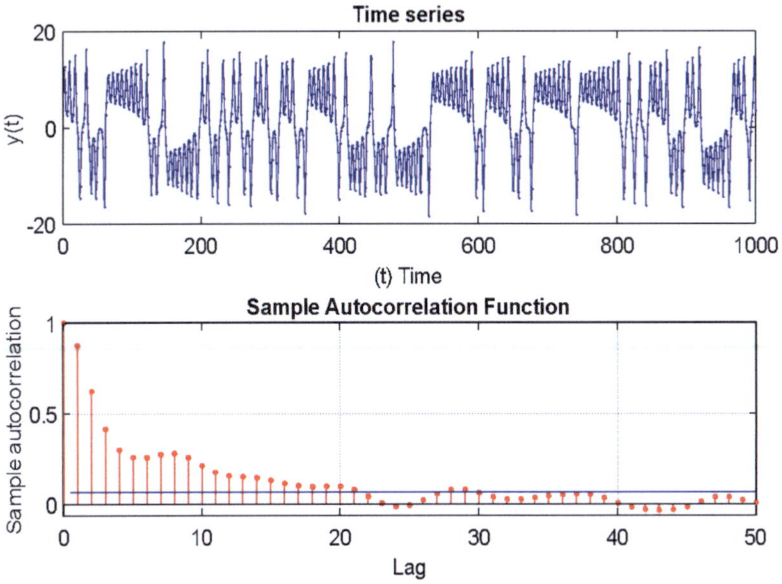

Fig. 2.8 Autocorrelation results of deterministic time series (Lorenz equations)

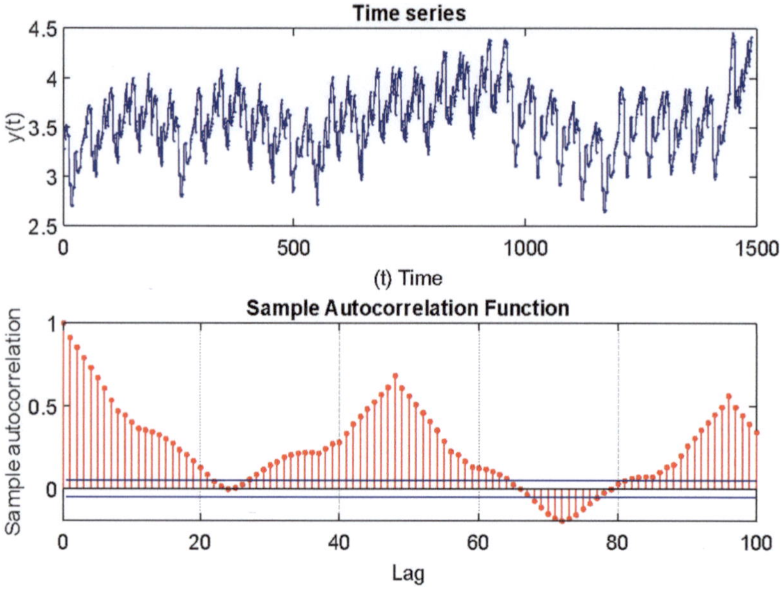

Fig. 2.9 Autocorrelation results of wind speed time series

2.2 Power Spectrum Analysis

Periodicity is closely related to the power spectrum. If a time series exhibits strong periodicity with a period T, its power spectrum will display a significant peak at the corresponding frequency $1/T$. Generally, a time series sampled at discrete time steps can be represented as a sum of periodic waveforms with different frequencies (or equivalently, different periods). This decomposition is expressed through the Fourier series, which takes the form:

$$x(t) = \alpha_0 + \sum_{k=1}^{M} \left(\alpha_k \cos\left(2\pi \mathrm{kf}t\right) + \beta_k \sin\left(2\pi ft\right) \right) \tag{2.2}$$

where α_0 is the mean, α_k and β_k are the amplitude for each cosine and sinus oscillation at harmonic frequencies, and M represents the number of harmonics, which can extend to infinity in a continuous case. f represents the fundamental frequency of the time series, which is the reciprocal of the fundamental period $T = 1/f$. Each term in the summation corresponds to a harmonic component of the signal, where kf denotes the k-th harmonic frequency (i.e., integer multiples of the fundamental frequency).

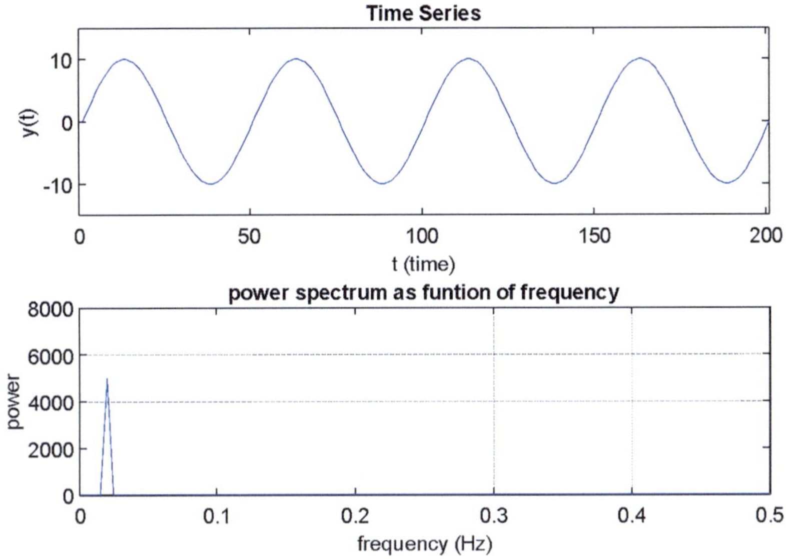

Fig. 2.10 Periodic time series and the power spectrum

If $x(t)$ is a periodic function with period T, then its Fourier transform consists of discrete frequency components, located at integer multiples of the fundamental frequency $f_0 = 1/T$. These components correspond to the coefficients in the Fourier series:

$$X(f) = \sum_{k=-\infty}^{\infty} c_k \delta(f - kf_0) \tag{2.3}$$

where c_k are the Fourier series coefficients, and $\delta(f)$ is the Dirac delta function, which indicates discrete frequencies.

The Fourier transform allows us to analyze how different frequency components contribute to the overall signal. A periodic signal has a discrete Fourier spectrum, meaning its Fourier transform consists of delta functions at discrete frequencies. A non-periodic (or transient) signal has a continuous Fourier spectrum, meaning it spreads across multiple frequencies.

Using script 2.2, we create a periodic time series and its Fourier transform, with the results appearing in Fig. 2.10. From the frequency chart, we can see that the time series has a main frequency $f = 0.02$ with a period $T = 1/f = 50$.

```
%%
%Script 2.2

function [ power ] = powerspectrum33(XV)

%UNTITLED2 Summary of this function goes here time series y1
%   Detailed explanation goes here

N=length(XV);
b=1:N;
Ts=1;
fs=1/Ts;
ts=Ts*(b-1);
X=fft(XV);
%X=fft(hanning(length(XV)).*XV)

pwr=X.*conj(X)/N ;
frs=(b-1)/N*fs;
% frs=(b-1)/N*fs;
subplot(2,1,1);
plot(XV),title('Time Series')
xlim([0 N])
xlabel('t (time)')
ylabel('y(t)')
subplot(2,1,2);
plot(frs,pwr),title('power spectrum as funtion of frequency')
grid on; xlabel('frequency (Hz)');ylabel ('power');
xlim([0 0.5])
[spow,spos]=sort(pwr);
m=4; spos(N:-1:(N-m+1));

end
```

Figure 2.11 below presents the frequency diagram of a time series resulting from a three-signal synthesis, and a different component is presented (script 2.2). From the frequency diagram, we can see these three different frequencies as well as the different intensity/contribution of each frequency. The first frequency is 0.02, the second 0.04, and the third 0.06. This illustrates the utility of the power spectrum analysis since it reflects both the frequencies present in the time series as well as their relative contributions.

Some points that should be taken into account when performing a power spectrum analysis that may affect the results.

Next we explore the so-called aliasing effect and the notion *of* Nyquist frequency. The Nyquist frequency is defined as half of the sampling frequency. In the following, we are going to see what the effect on a recorded signal is if the Nyquist frequency is higher than the largest frequency of the system under study and how the signal is distorted and as a consequence the Fourier transforms too.

In 2.12, we have a periodic signal with frequency 4 Hz and we present with circles the points recorded at various sapling frequencies and specifically Fs = 20 Hz,

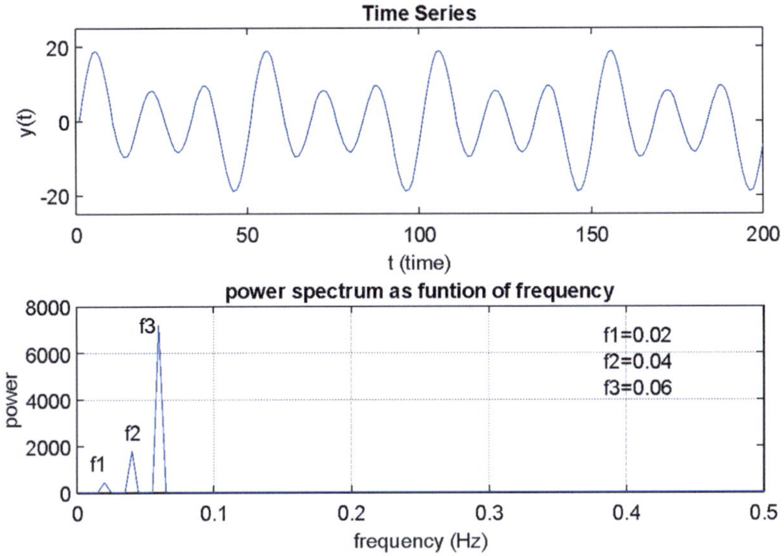

Fig. 2.11 Periodic time series consist of three frequencies and the corresponding power spectrum

10 Hz, and 6 Hz. This means that the corresponding Nyquist frequencies FNyquist = 10 Hz, 5 Hz, and 3 Hz, respectively. We can see that for the case of FNyquist<Fsystem, the signal is completely distorted and seems to correspond to a lower frequency. So when the distorted signal is Fourier transformed, it will result in lower frequency.

Initially, using script 2.3, we create the initial time series of Fig. 2.12.

```
%Script 2.3

% equation  y(t) = sin(2*pi*f*t)

f=4;      % frequency
Ts=0.01; % sampling rate fs=1/Ts (100/sec)
t=0: Ts: 1;
x=sin(2*pi*f*t);

plot(t,x, 'o-');

title('Initial Time series','FontSize',20)
ylabel('y(t)')
xlabel('(t) Time')
```

Next, in order to see how the resampling affects the time series, we use the following script 2.4 in Matlab.

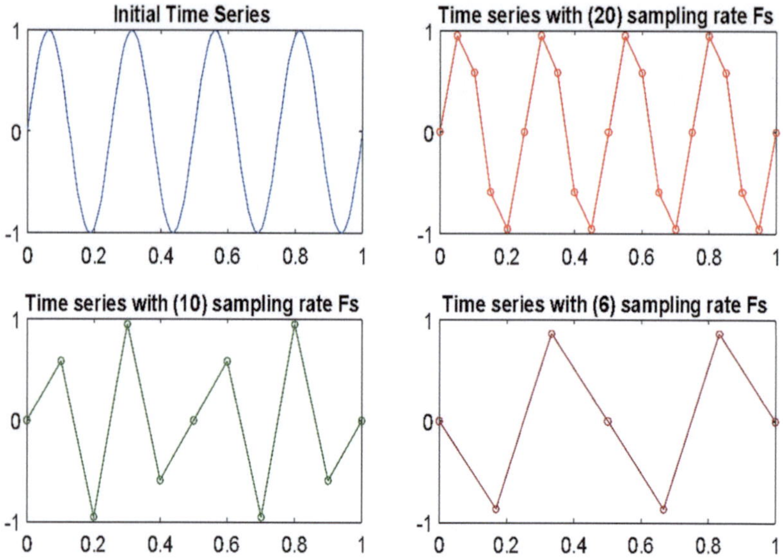

Fig. 2.12 Periodic time series and the extracted time series with different sampling rate

```
%%
%Script 2.4

% equation  y(t) = sin(2*pi*f*t)
f=4;
Fs=input('sampling frequency_');
Ts=1/Fs;
t=0:Ts:1;
[x1]=(sin(2*pi*f*t))';
plot(t,x1, 'o-');
title(sprintf('Time series with (%d) sampling rate Fs
',Fs),'FontSize',14)
```

The Nyquist frequency is calculated as $2 \cdot f = 2 \cdot 4 = 8$.

Subsequently, the results are presented in comparable figures (Fig. 2.13).

Then, a periodic time series with three frequencies is constructed, where different cases are given depending on the sampling frequency (Figs. 2.14 and 2.15).

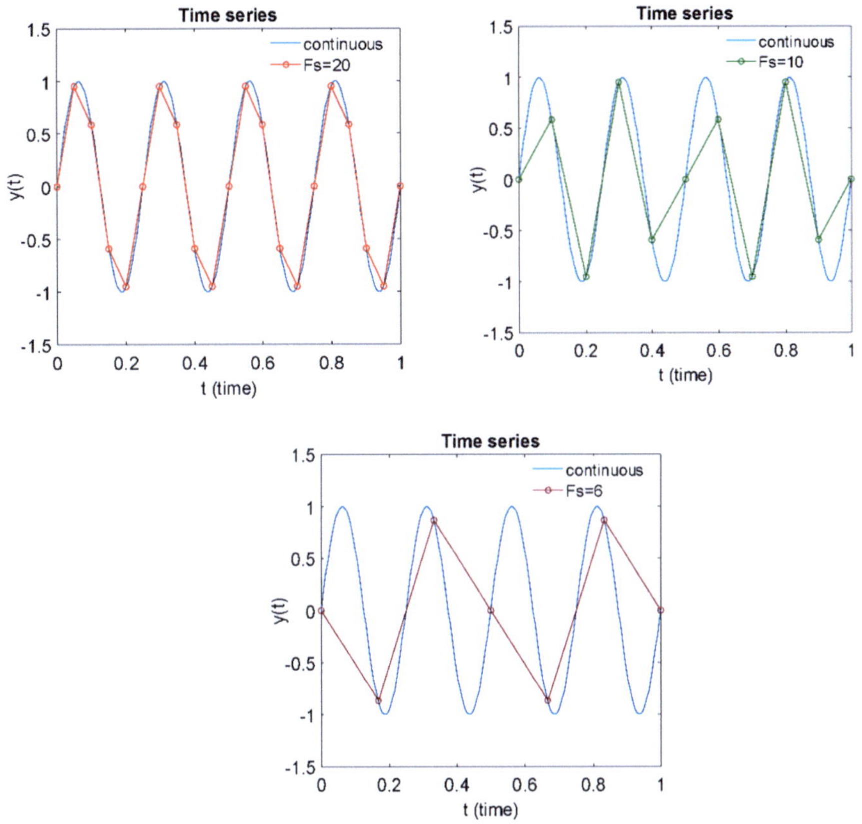

Fig. 2.13 Effect of sampling rate on time series recording

```
%%
%Script 2.5

% equation
%y(t)=5*cos(2*f1*pi*t)+15*cos(2*f2*pi*t)+5*cos(2*f3*pi*t);

Fs=input('sampling frequency_');    %400
f1=input('frequency_1_');     %5
f2=input('frequency_2_');     %10
f3=input('frequency_3_');     %15
Ts=1/Fs;
t=0:Ts:0.4;
x=5*cos(2*f1*pi*t)+15*cos(2*f2*pi*t)+5*cos(2*f3*pi*t);
plot(t, x, 'o-');
title(sprintf('Time series with (%d) sampling rate Fs
',Fs),'FontSize',14)
ylabel('y(t)')
xlabel('t (time)')
```

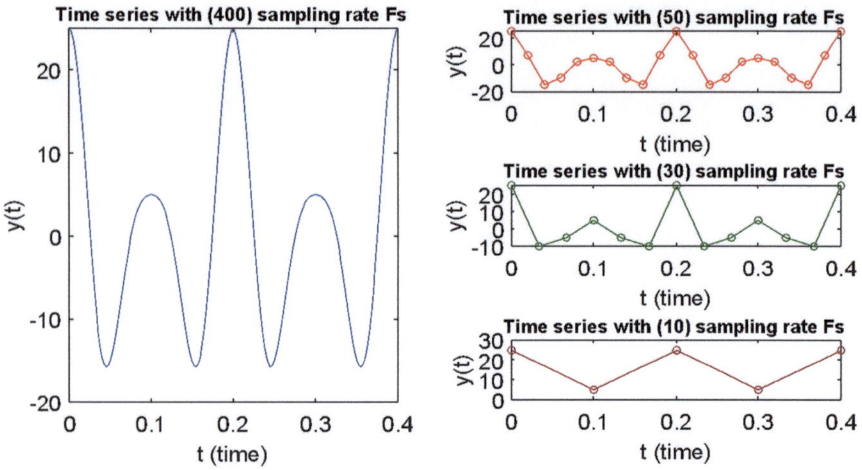

Fig. 2.14 Time series and the extracted time series with different sampling rate

From the above cases, we can see the effect of sampling frequency on different time series cases.

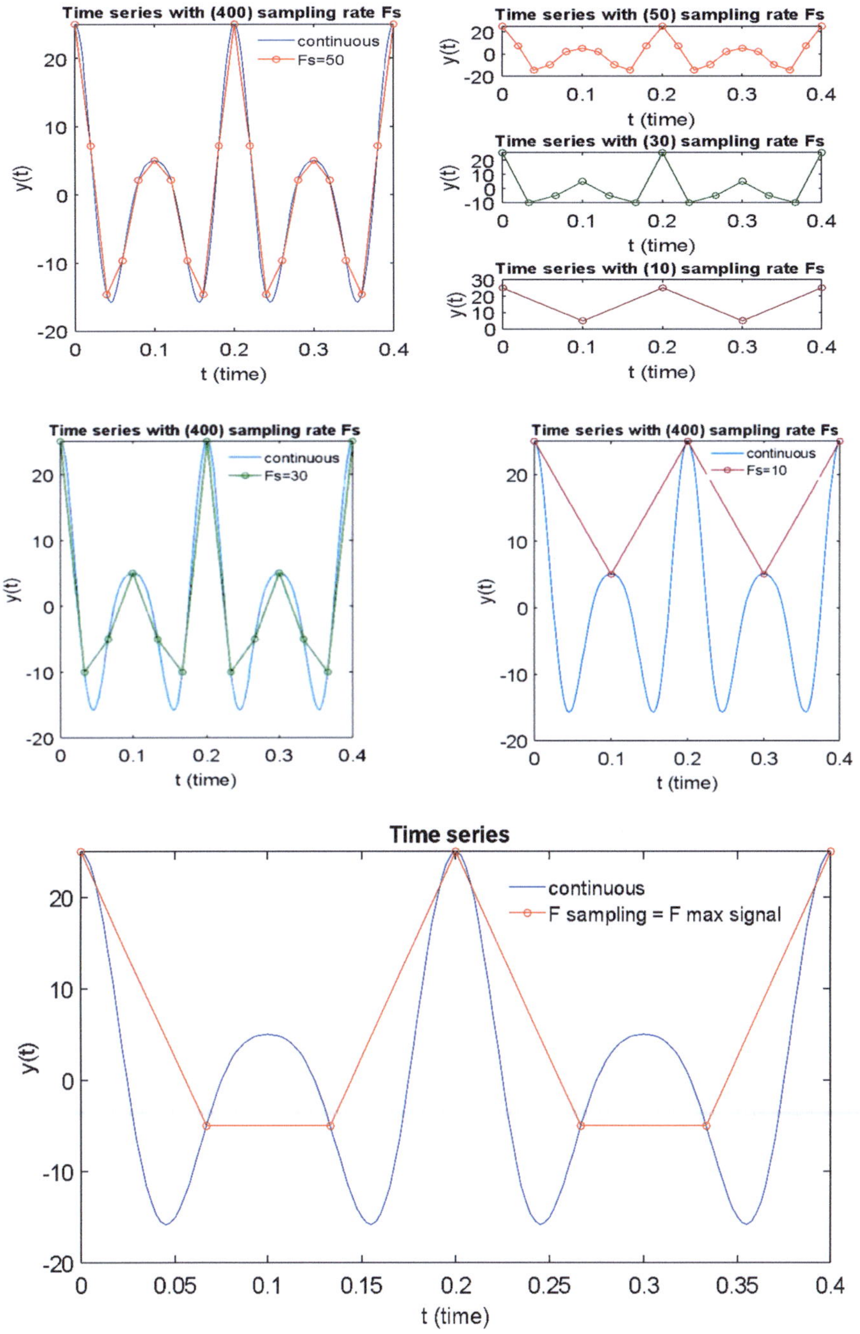

Fig. 2.15 Continuous time series and the extracted time series with different sampling rate and a time series with sampling rate equal to F maximum

2.3 Mutual Information

Mutual information often proves to be statistically useful, especially when it comes to assessing the association between two variables in time series analysis. Unlike the correlation coefficient, which only indicates the strength of linear relationship, mutual information can find both linear and non-linear relationships between the variables [3, 5].

Mutual Information $I(t)$ is a widely used nonlinear measure used in time series analysis for determine the appropriate delay time τ for state space reconstruction and is defined as:

$$I(t) = \sum_{x(\mathrm{ti}),x(\mathrm{ti}+t)} P(x(\mathrm{ti}), x(\mathrm{ti}+t)) \log\left[\frac{P(x(\mathrm{ti}),x(\mathrm{ti}+t))}{P(x(\mathrm{ti}))\,P(x(\mathrm{ti}+t))}\right] \tag{2.4}$$

where $x(\mathrm{ti})$ is the ith data point of time series, $t = k\Delta t$ ($k = 1,2,\ldots\, k_{\max}$); Δt is the sampling time; $P(x(\mathrm{ti}))$ is the probability density at $x(\mathrm{ti})$, $P(x(\mathrm{ti}), x(\mathrm{ti} + \tau))$ is the joint probability density at $x(\mathrm{ti})$, $x(\mathrm{ti} + t)$; τ is the delay time.

The delay t corresponding to the first minimum of the mutual information is chosen as a delay time for the reconstruction of phase space.

Using script 2.6, we produce a periodic signal that appears in Fig. 2.16, and then we run the mutual information script to obtain the results in Fig. 2.17.

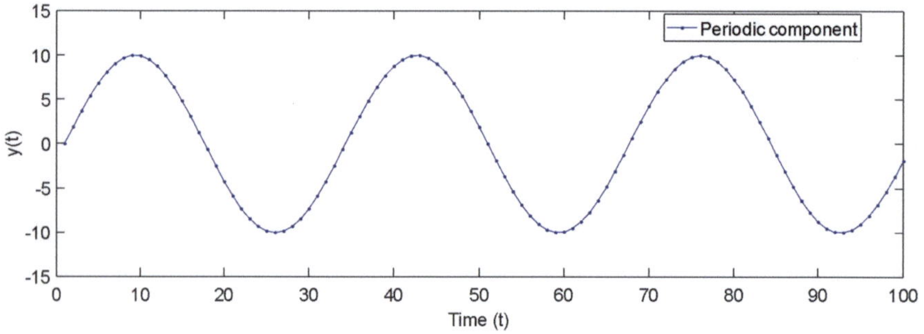

Fig. 2.16 Periodic time series

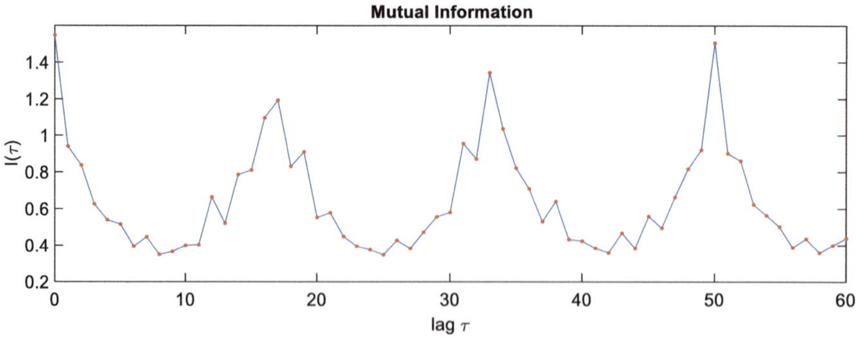

Fig. 2.17 Results of mutual information function with time lag $\tau = 60$

```
%%
%Script 2.6
%Generate periodic time series

N=input('Give the time series length_:');
f=input('Give the frequency_:');
Amp=input('Give the Amplitude_:'); 100,3,10
t=0:1:N;
yy=Amp*sin(2*pi*f*t/100);
plot(yy,'b.-','MarkerSize',10);    %plot function
ylim([-Amp-5 Amp+5])
xlim([0 N])
xlabel('Time (t)');
ylabel('y(t)');
legend('Periodic component ','Location','Best')
```

In Fig 2.17 we applied the mutual information function for a delay τ=60 using the following command, and where we can see the existence of periodicity, which is a result of the periodic nature of the time series.

```
%%
%Mutual information

M=mutualinformation(yy,60)
```

The next figure shows the result of mutual information function, where we can see the existence of periodicity which is a result of the periodicity of the time series.

Then we calculate the mutual information for a shorter time (Fig. 2.18).

```
%%
%Mutual information

M=mutualinformation(yy,10)
```

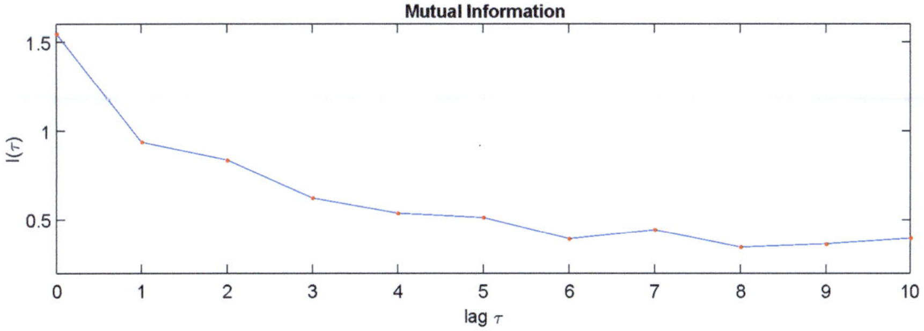

Fig. 2.18 Results of mutual information function with time lag $\tau = 10$

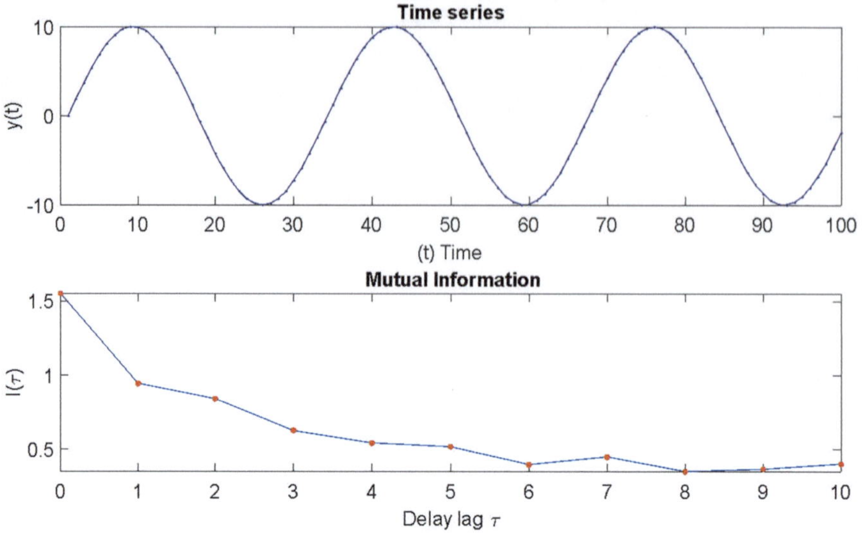

Fig. 2.19 Time series and the correspondance mutual information function

In the following script 2.7, we give as input the time series name and the maximum delay time to be explored and we obtain a plot of the mutual information diagram that corresponds to the time series (Fig. 2.19).

```
%%
% script 2.7
data=input('Time series name_');
tmax=input('Time Lag_');
figure
subplot(2,1,1);
plot(data,'b.-','MarkerSize',6);
axis([0 100 -10 10])
title('Time series','FontSize',10)
ylabel('y(t)')
xlabel('(t) Time')
subplot(2,1,2);
[mutM] = mutualinformation_b(data, tmax)
```

So far, we have seen the results of the function for synthetic time series. Then the result for field measurements and specifically for the wind velocity time series, the results are obtained running the following command and appear in Fig. 2.20.

```
%%

%Mutual information

M=mutualinformation(wind,100)
```

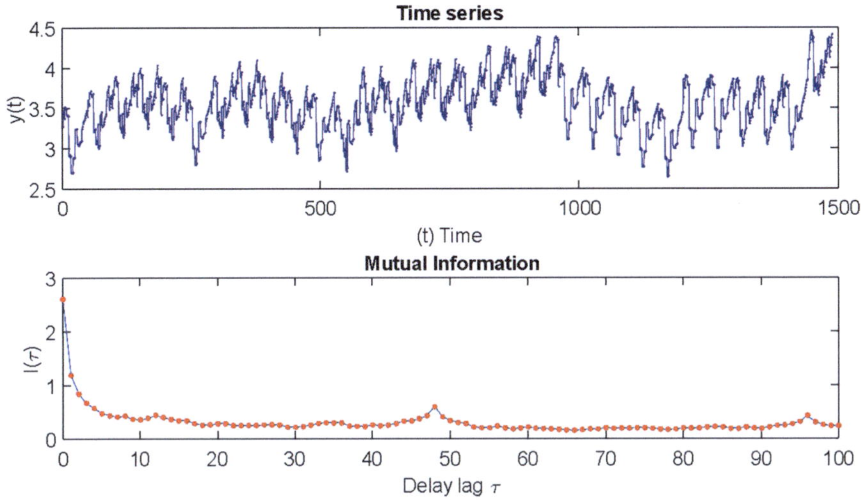

Fig. 2.20 Results of mutual information function with time lag $\tau = 100$ of wind speed time series

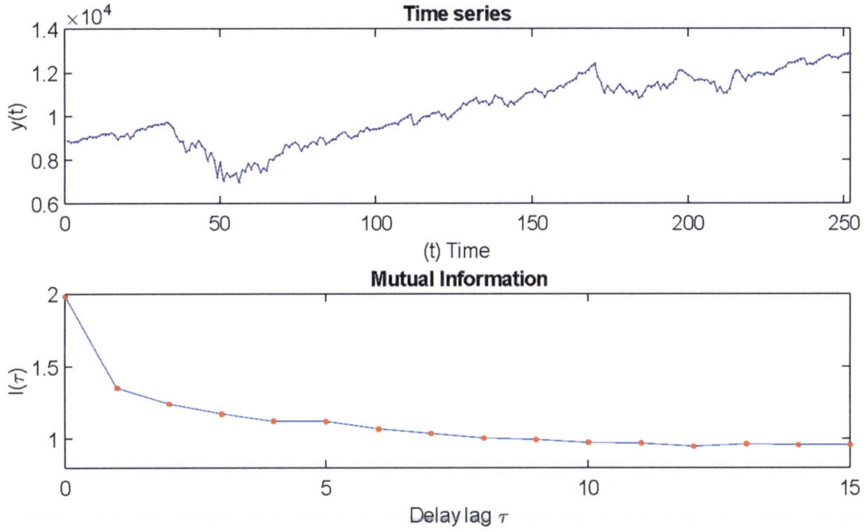

Fig. 2.21 Results of mutual information function with time lag $\tau = 100$ of Nasdaq time series

The results for time series of stock market prices are also presented in Fig. 2.21.

Mutual information is particularly valuable in phase space reconstruction methodology, as it helps determine the optimal time delay between points in the reconstructed space by identifying the delay that maximizes the independence between successive points. This step is essential for accurately capturing the system's dynamics and reducing redundancy in the reconstructed phase space, ultimately leading to a clearer and more informative representation of the system's behavior.

2.4 Hurst Exponent

The measurement of long-term memory in time series is done through the estimation methodologies of the Hurst exponent [2]. Hurst created Rescaled Range (R/S) analysis which is one of the most widely used methods for estimating the Hurst exponent. In Rescaled Range (R/S) analysis, we start by splitting the time series $\{x_i\}.i = 1,\ldots N$ of length N to S shorter time intervals of length $n = N, N/2, N/4$. Then, for each time interval (time series), we calculate the range R_n.

$$R_n = \max_{1 \le k \le s}\left[\sum_{i}^{k}\left(x_{\mathrm{ns}+i} - \overline{x_n}\right)\right] - \min_{1 \le k \le s}\left[\sum_{i}^{k}\left(x_{\mathrm{ns}+i} - \overline{x_n}\right)\right] \tag{2.5}$$

where $n = 0, 1, \ldots\ldots, N_s - 1$ με $N_s = N/S$ and

$$\overline{x_n} = \frac{1}{s}\sum_{i=1}^{s} x_{\mathrm{ns}+i} \tag{2.6}$$

The standard deviation of time series is given by:

$$S_n = \sqrt{\frac{1}{s}\sum_{i=1}^{s}\left(x_{\mathrm{ns}+i} - \overline{x_n}\right)^2} \tag{2.7}$$

The rescale range R/S is defined as the average R_n/S_n ratio of all time intervals. In other words, it is given by:

$$R/S = E\left[\frac{R_n}{S_n}\right] \tag{2.8}$$

The Hurst exponent is calculated from the scaling behavior of the rescalable range R/S [7].

$$\left(R/S\right)_n = c \cdot n^H \qquad s \to \infty \tag{2.9}$$

In order to calculate the exponent, the logarithm graph of the mean rescaled amplitude $(R/S)_n$ versus the length n is created. Then, a linear regression line is adjusted to the graph, the slope of which gives the estimate of the Hurst exponent.

Observing that the calculation of the Hurst exponent using the R/S method led to erroneous conclusions regarding the existence of large-scale correlations, the method of Detrended Fluctuation Analysis (DFA) was developed and applied. This method is a version of the initial variance analysis, in which linear trends are eliminated from the time series and is applied in cases of non-stationary time series.

Initially, for a time series of length N, the sum or profile is calculated:

$$Y(i) = \sum_{k=1}^{i}\left[x_k - \overline{x}\right] \tag{2.10}$$

The time series $\{x_i\}$ $i = 1,\ldots N$ of length N is divided to S shorter time intervals of length $n = N, N/2, N/4,\ldots$

Then, in each section, the polynomial of degree m is estimated by adjusting an appropriate polynomial. Trend-free time series are defined as the difference between original time series and adaptations:

$$Y_n(i) = Y(i) - Y_{v,n}^m(i) \tag{2.11}$$

where $Y_{v,n}^m(i)$ it is the polynomial that adapts to the N-th segment.

For each segment, the variance is calculated

$$F_v^2(n) = \frac{1}{n}\sum_{i=1}^{n} Y_n^2(i) \tag{2.12}$$

Finally, calculating the square root of the mean of the variances in each segment estimates the DFA function

$$F_{\mathrm{DFA}}(n) = \left[\frac{1}{2N_n}\sum_{v=0}^{2N_{s-1}} F_n^2(n)\right] \tag{2.13}$$

Hence, we can plot the diagram $\log(R/S)_n$ vs $\log(n)$ and we calculate the Hurst exponent using linear least squares regression.

In the following, we run the Matlab command window, for the wind time series (Fig. 2.22).

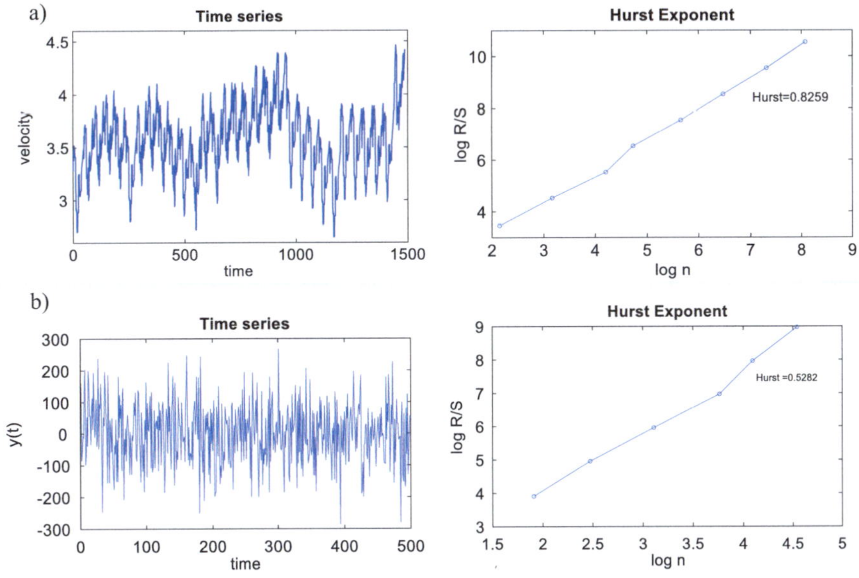

Fig. 2.22 Time series of wind speed (**a**) and random time series (**b**) and the corresponding result of Hurst exponent

```
%%
%Hurst Exponent

H= HurstExponent(wind)
```

The Hurst exponent takes values within the closed interval [0,1]. Its interpretation depends on how it compares to 0.5. When the exponent H is equal to 0.5, then the time series follows the random walk model, which means that there is no correlation between the values. In this case, the measurements are independent of each other. Conversely, when the exponent is different from zero, it means that the observations are not distributed independently but have a long-term memory.

According to this definition, a Hurst exponent of 0.5 indicates a purely random process, where past values have no influence on future values.

- When $0.5 < H < 1$, the time series exhibits persistent behavior, meaning that high values tend to be followed by high values and low values by low values. This suggests a long-term positive correlation.
- When $0 < H < 0.5$, the time series demonstrates anti-persistent behavior, where high values are more likely to be followed by low values and vice versa. This indicates a tendency to revert to the mean overtime.

2.5 Hjorth Parameters

Hjorth parameters are characteristic measures of time series and are used to quantitatively describe a time series [1]. For a time series $x(t)$, the following parameters shall be set:

$$\text{Activity} = m_o \tag{2.14}$$

$$\text{Mobility} = \sqrt{\frac{m_2}{m_0}} \tag{2.15}$$

$$\text{Complexity} = \sqrt{\frac{m_4 / m_2}{m_2 / m_0}} \tag{2.16}$$

where m_0 is the variance (square of the standard deviation) of the variable, m_2 is the variance of the first derivative of the variable, and m_4 is the variance of the second derivative of the variable.

The mobility parameter expresses the average frequency of the time series and is calculated as the ratio at each point in time of the standard deviation of the time series slope to the standard deviation of the time series. The complexity parameter represents the change in the frequency of the time series and is defined as the ratio of (a) the ratio of the second-order central moment of the second derivative to the second-order central moment of the first derivative and (b) the ratio of the second-order central moment of the first derivative to the second-order central moment of the original time series. Complexity expresses the deviation of the slope and can be thought of as a measure of the change in the frequency of the input signal.

Using the Matlab function below, we obtain the Hjorth parameters in the case of wind times series and a reando times series (see Fig. 2.23). It is clear that the activity in the case of time series is significantly larger than in the wind time series.

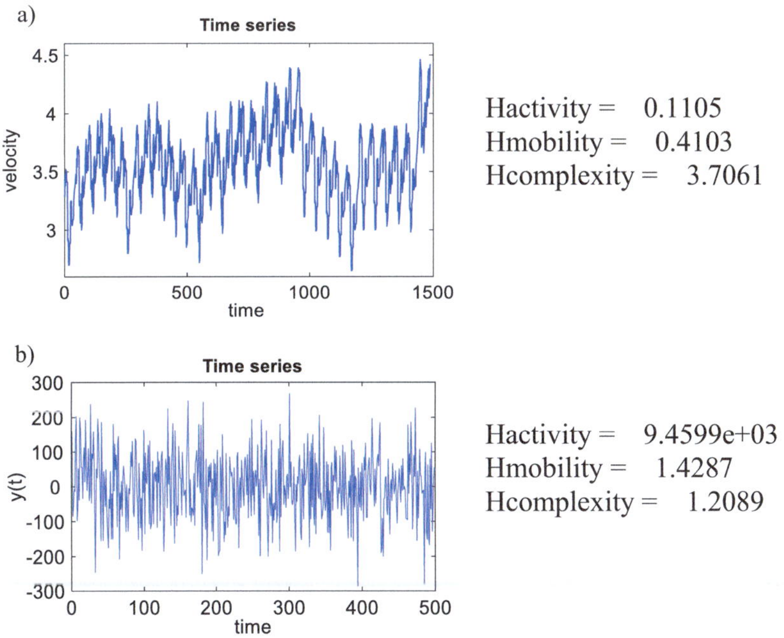

Fig. 2.23 Time series of wind speed (**a**) and random time series (**b**) and the corresponding result of Hjorth parameters

```
function [Hactivity,Hmobility,Hcomplexity] =
HjorthParameters_book(TS)
% [Hmobility,Hcomplexity] = HjorthParameters(TS)
% estimate the Hjorth  mobility and complexity.
% INPUTS:
% - TS            : The time series
% OUTPUTS
% - Hactivity,Hmobility,Hcomplexity
%==================================================================
=======

dTS = diff(TS);
ddTS = diff(dTS);
m0 = var(TS);
m1 = var(dTS);
m2 = var(ddTS);

Hactivity = m0
Hmobility = sqrt(m1/m0)
Hcomplexity=sqrt((m2/m1)/(m1/m0))
```

2.6 Clustering

Cluster analysis or data clustering is characterized as the partitioning of a set of raw data into subsets, resulting in the extraction of useful information from them. It is extensively used in scientific research across various fields such as medicine, biology, statistics, and engineering problems. The goal of clustering is to identify structures within a dataset or, more simply, to create groups where each group gathers homogeneous elements based on some similarity measure. The efficiency of clustering methods is directly related to the type of data as well as the homogeneity criterion or similarity measure used.

There are various clustering techniques, and their selection depends on the nature of the data and the purpose of the clustering. One of the most common approaches is hierarchical clustering [6]. Hierarchical clustering is performed in two ways: agglomerative analysis and divisive analysis. In agglomerative analysis, initially, each data point is considered as a separate entity. Then, at each iterative step, elements are merged based on a specific criterion, forming larger groups until all elements ultimately form a single group. This method requires a similarity criterion and a proximity measure, which is defined as the distance between two elements. In contrast, in divisive analysis, all data elements initially belong to a single group (a unified cluster). Subsequently, based on a specific criterion, the elements are successively divided into smaller groups until each data element becomes its own separate cluster, creating as many groups as there are data points.

The result of hierarchical clustering is a tree of groups and connections called a dendrogram, which illustrates how the groups are related to one another. In the dendrogram, a horizontal cut is chosen at a specific level. At this point, the number of resulting groups and the elements contained in each group are displayed. Figure 2.24

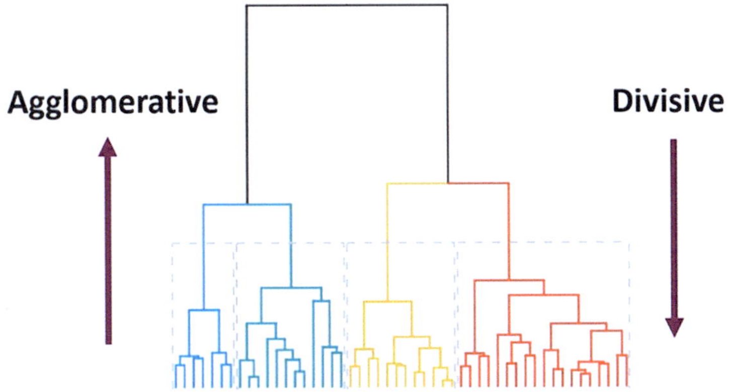

Fig. 2.24 Example of hierarchical clustering

Fig. 2.25 Single-linkage clustering

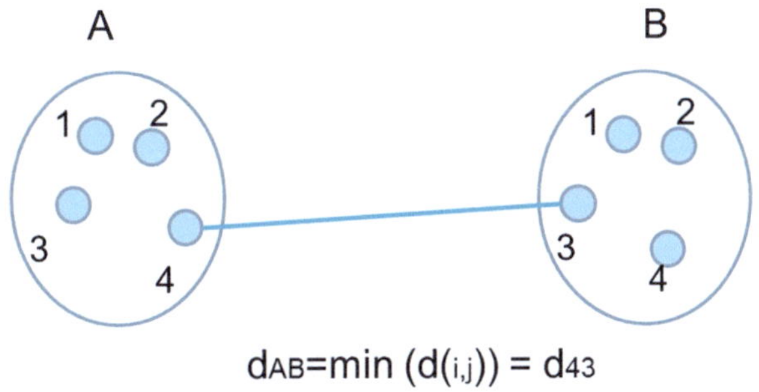

presents a typical dendrogram, where the horizontal axis refers to the groups (as many as the data points), and the vertical axis corresponds to the level of similarity or dissimilarity. By making horizontal cuts in the dendrogram, we can observe the number of groups at the similarity level of interest.

There are various criteria used for dividing data into groups, which are based on the distance matrix between pairs of data points. In other words, the similarity or dissimilarity between the data is measured based on a distance function between the elements or groups.

2.6.1 *Single-Linkage Clustering or Nearest Neighbor*

The single-linkage or nearest neighbor method uses the minimum distance between elements, and subsequently between groups, as the similarity criterion. According to this method, the two closest elements are initially connected based on the smallest distance, and in each iterative step, distances are recalculated, and connections are gradually made until a single group containing all elements is formed. The algorithm is illustrated schematically in Fig. 2.25.

2.6.2 Complete-Linkage Clustering

The process of this method differs from single-linkage in that complete-linkage clustering uses the maximum distance between elements or groups. Figure 2.26 graphically illustrates the algorithm.

2.6.3 Average-Linkage Clustering

In hierarchical average-linkage clustering, the distance between two groups is defined as the average distance between all possible pairs of elements from each group. The weighted average distance is calculated based on the number of data points in each group. Figure 2.27 illustrates the connections and the formula for calculating the average distance.

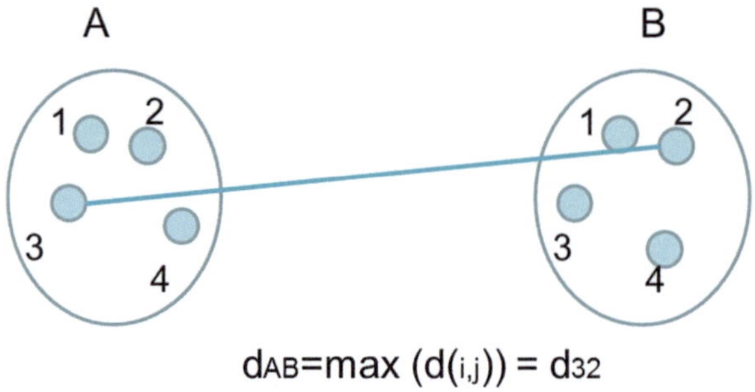

Fig. 2.26 Complete-linkage clustering

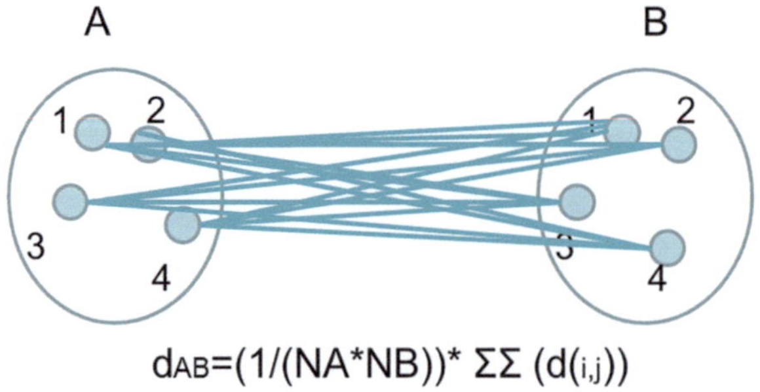

Fig. 2.27 Average-linkage clustering

2.6.4 Centroid-Linkage Clustering

In this method, the Euclidean distance between the centroids of the groups is used as the distance criterion. In each step of the algorithm, two groups are merged based on the smallest distance between their centroids. Figure 2.28 graphically depicts the algorithm.

These methods were applied to cluster the time series data to examine whether clustering could provide better insights into the separation of regions involved in the experimental process of turbulent flow. Notably, the clustering methodology did not use the same time series as a similarity measure but rather a vector consisting of linear and nonlinear measures, as extensively discussed earlier.

The dendrogram below illustrates the clustering of time series, using the "Euclidean" metric for calculating the distance matrix among all elements and the single-linkage method, as it achieves the highest cophenetic correlation coefficient (Fig. 2.29).

In cases where the data are either too large or contain discontinuities, instead of using the raw time series data for clustering, we can compute the descriptive measures of the time series and then input these into the clustering algorithm. This approach reduces the length of the input vector, allowing the routine to produce results much faster.

In following, the statistical descriptive measures of the time series and calculate the dendrogram based on these measures and not using the time series values (Fig. 2.30).

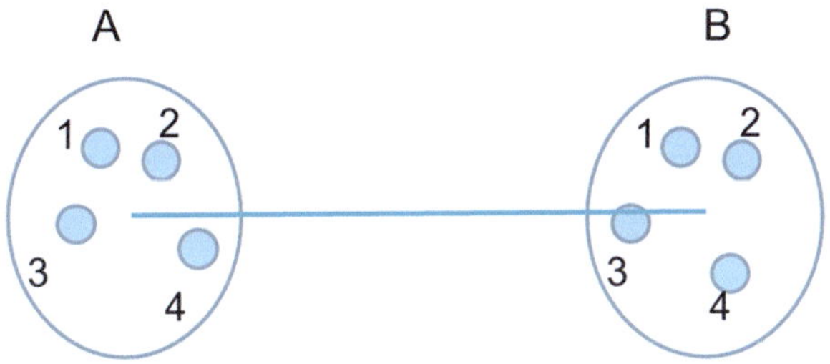

Fig. 2.28 Centroid-linkage clustering

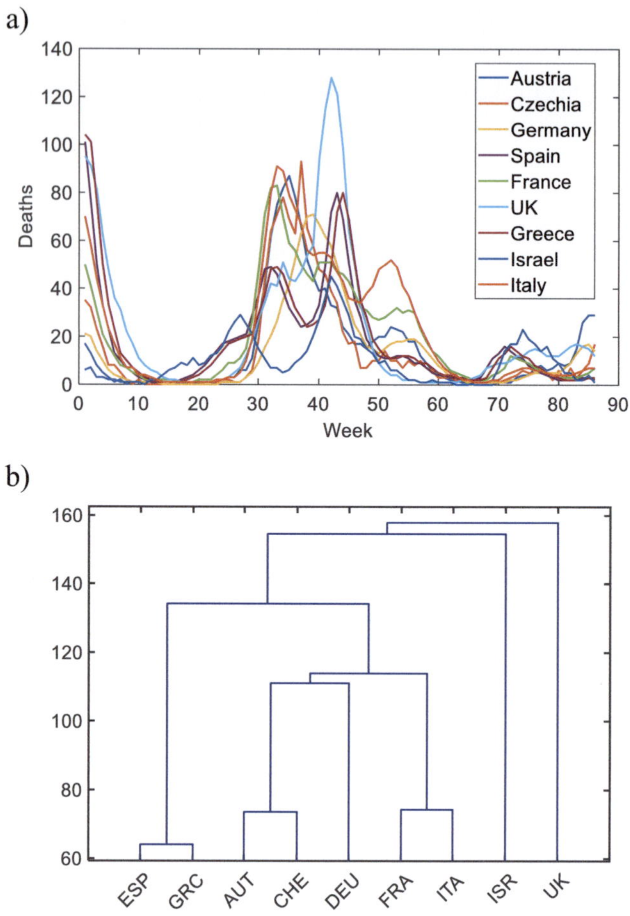

Fig. 2.29 (**a**, **b**) Time series of COVID deaths (weekly) of countries and the chart of hierarchical clustering (https://www.worldometers.info/coronavirus/)

Fig. 2.30 Hierarchical clustering based on statistical measures of time series values

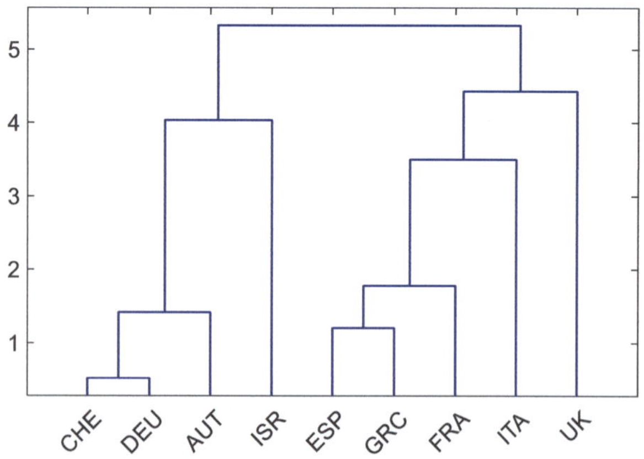

References

1. Hjorth, B. (1970). EEG analysis based on time domain properties. *Electroencephalography and Clinical Neurophysiology, 29*(3), 306–310.
2. Hurst, H. E. (1951). Long-term storage capacity of reservoirs. *Transactions of the American Society of Civil Engineers, 116*(1), 770–799.
3. Kantz, H., & Schreiber, T. (2003). *Nonlinear time series analysis*. Cambridge University Press.
4. Koopmans, L. H. (1995). *The spectral analysis of time series*. Elsevier.
5. Kraskov, A., Stögbauer, H., & Grassberger, P. (2004). Estimating mutual information. *Physical Review E—Statistical, Nonlinear, and Soft Matter Physics, 69*(6), 066138.
6. Kraskov, A., Stögbauer, H., Andrzejak, R. G., & Grassberger, P. (2005). Hierarchical clustering using mutual information. *Europhysics Letters, 70*(2), 278.
7. Mandelbrot, B. B., & Wallis, J. R. (1969). Robustness of the rescaled range R/S in the measurement of noncyclic long run statistical dependence. *Water Resources Research, 5*(5), 967–988.

Chapter 3
Nonlinear Time Series Analysis

Nonlinear time series analysis plays a vital role in understanding the complex behaviors exhibited by dynamical systems over time. Unlike linear methodologies, which assume straightforward relationships between variables, nonlinear analysis delves into the intricate interdependencies and feedback loops that characterize many real-world phenomena. By examining the underlying dynamics across diverse fields such as meteorology, economics, neuroscience, and ecology, nonlinear time series analysis unveils emergent patterns, chaotic behavior, and predictive insights that often elude traditional linear approaches. Through sophisticated mathematical techniques like phase space reconstruction, chaos theory, and recurrence quantification analysis, researchers can capture the nonlinear dynamics inherent in these systems, offering a deeper understanding of their underlying mechanisms and behaviors [1, 2, 6, 16].

In recent years, the application of nonlinear time series analysis has gained ground due to advances in computational power and the recognition of its relevance across an expanding array of disciplines. From financial forecasting to climate modeling, from biological systems to engineering applications, nonlinear analysis has proven indispensable in deciphering the complexities of dynamic systems. As researchers continue to refine methodologies and develop novel techniques, nonlinear time series analysis remains at the forefront of scientific inquiry, providing invaluable tools for exploring the rich tapestry of nonlinear dynamics that underpin our ever-evolving world.

3.1 Introduction to Dynamical System

A dynamical system is defined as any system that evolves over time. In a dynamical system, there exists a set of variables that interact with each other, generating the system's behavior, which corresponds to the system's variables. Additionally, in a system, the term "state" is defined as the set of variables $x_1\,(t)$, $x_2\,(t)$, …,$x_n\,(t)$ that

T. Karakasidis, A. Charakopoulos, *Time Series and Networks Analysis*,
https://doi.org/10.1007/978-3-031-92628-0_3

describes the system's state at the time t. The set of successive states defines the so-called state space of the dynamical system. Otherwise stated, the system's state space is the set of possible positions-solutions of the system, where the system is represented as a function of its variables, connecting the past value with the present and future. The dimension of the state space is defined by the number of variables needed to describe the system.

The mathematical definition of a dynamical system states that a dynamical system is any system whose evolution from some initial state is mathematically described by a system of differential equations, where the independent variable is time (t).

We assume that the time series $x_{(i)}$ originates from a system of n differential equations describing the evolution of the system.

$$
\begin{aligned}
\frac{dx_1}{dt} &= f_1\left(x_1, x_2 \ldots, x_n; c_1, c_2, \ldots, c_n\right),\\
\frac{dx_2}{dt} &= f_2\left(x_1, x_2 \ldots, x_n; c_1, c_2, \ldots, c_n\right),\\
&\cdot\\
&\cdot\\
\frac{dx_n}{dt} &= f_n\left(x_1, x_2 \ldots, x_n; c_1, c_2, \ldots, c_n\right),
\end{aligned}
\tag{3.1}
$$

or equivalently

$$
\frac{d\boldsymbol{x}}{dt} = \boldsymbol{F}\left(x, t, c\right) \tag{3.2}
$$

where $\boldsymbol{x}$ is a vector of n time variable

$$
x(t) = \left[x_1(t), x_2(t), \ldots x_n(t)\right] \tag{3.3}
$$

In Eq. (3.1), F is the function describing the system with F: $R^d \rightarrow R^d$, where R^d is the state space or phase space of the system, d is the dimension of the Euclidean state space, and c represents the parameters of the system that remain constant. The dimension of the state space is denoted by n, equal to the number of variables. At each moment in time, the position of the system in the state space is given by the term.

The variables x_i typically represent physical quantities such as position, velocity, temperature, pressure, etc. The state of the system at a specific moment in time in phase space is depicted by a point $\boldsymbol{x}(t) = [x_1(t), x_2(t), \ldots x_n(t)]$. At the next moment in time, the state of the system will change, and it will move to a new state point, creating in this way a trajectory after successive time steps in phase space, which illustrates the temporal evolution of the dynamical state of the system. Each point in phase space traces only one trajectory due to the deterministic component of the system. However, changing the initial conditions or the parameters of the system can alter the behavior of the dynamical system.

Dynamical systems can be classified as deterministic or stochastic, depending on whether their behavior is predictable. A system is deterministic if its state at a given time uniquely determines all future states. In such systems, both past and future states are completely specified by the governing equations. In contrast, stochastic dynamical systems are influenced by random external or internal factors, making their future evolution inherently uncertain at least after a given time.

A particularly important subclass of deterministic systems consists of chaotic systems. A dynamical system exhibits chaos when small variations in initial conditions lead to exponentially diverging trajectories, resulting in vastly different long-term outcomes. Despite being deterministic, chaotic systems are highly sensitive to initial conditions, making long-term prediction practically impossible.

Additionally, dynamical systems are classified based on how they evolve over time. If a system is described by differential equations, it is a continuous-time system. If it is governed by difference equations, it is a discrete-time system.

In the present book, we are going to deal with continuous and chaotic systems.

3.1.1 System Identification

When the evolutionary equations describing it are known for a dynamical system, then we can generate the time series from the equations that govern the system. So, suppose the state of a system at time t is x_t, and the difference equation describing it is of the form $x_t + 1 = f(x_t)$, if we know the initial state x_0 and the function f describing the dynamic behavior of the system, we can recursively compute the state of the system for each time step. The ability to generate data translates into a complete understanding of the system's dynamic behavior. In other words, in this way, we have knowledge of the past, the current state, and we can predict the future states of the system.

However, the study of physical dynamical systems suggests that purely stochastic or purely deterministic behavior is the exception rather than the rule. In contrast, studying a fully deterministic system is relatively straightforward, as each state follows directly from the previous one, and the next state is uniquely determined by the current one or some previous ones. However, in many physical but also economic or biological systems, determinism and randomness coexist, leading to the notion of complexity. Complexity may also arise from the large number of variables that may determine the state of a system. For example, a system composed of many interacting atoms may have well-defined interaction rules, yet its global behavior remains difficult to predict due to emergent complexities.

Physical and other category dynamical systems generally arise from nonlinear equations involving multiple variables, the exact analytical form of which is often unknown. As a result, interpreting these systems is highly challenging. Their study relies on time series analysis. A time series, whether in single-variable or multivariable analysis, represents recorded observations of a system's evolution over time and often provides the only available insight into its dynamics. Time series are intrinsically linked to the system's dynamic behavior.

In time series analysis, the term "system identification" refers to a set of methods designed to understand the temporal evolution of data. The primary goal is to analyze the system's behavior, extract the underlying dynamics, and identify representative characteristics, rather than fully uncovering the internal mechanisms or physical laws governing the process. Additionally, system identifications seek to determine the key elements influencing the system's evolution, namely its dynamics, using only measured data. These measurements must be strongly coupled with the system's dynamics to ensure accurate modeling and interpretation.

3.1.2 Phase Space Reconstruction

As previously mentioned, the dynamic behavior of a system in many cases can be described by differential equations and refers to the temporal evolution of the system's state [3–5, 14–17]. The description of the system is determined by a number of n variables over time, which determine the state of the system, i.e., they constitute the "solution" of the system for a given moment in time. The space defined by the number of these dynamic variables is called the state space or phase space, which has a dimension equal to the number of variables and is the space where the position state of the system is depicted for each moment in time. The solution of the dynamical system, for specific initial conditions, in phase space is represented by the successive positions of the state vector or alternatively by the solutions of the system. These successive solutions form the trajectory or curve of the system. Using the script below (script 3.1), we produce representative time series for a harmonic oscillator with a given frequency and amplitude and then we construct the corresponding trajectory in the phase space. The corresponding results appear in Fig. 3.1 where we

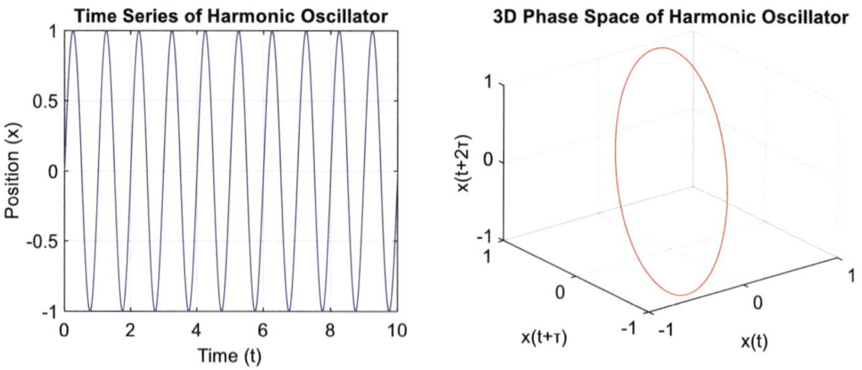

Fig. 3.1 Time series of harmonic oscillator and the phase space, respectively

can see that the time series of the harmonic oscillator with constant amplitude gives us a circle which is an attractor of the system. The interesting part is that knowing this circle we know all the states that the system can visit.

```
% script 3.1
%Time series for a simple harmonic oscillator
t = 0:0.01:10;                  % Vector (0 to 10 seconds with 0.01s
intervals)
omega = 2*pi;                   % Angular frequency (1 Hz)
x = sin(omega*t);               % Position of the harmonic oscillator

% Choose the time delay (tau)
tau = 10;   % Example time delay (adjustable)

% Create the 3D phase-space (state space reconstruction)
X1 = x(1:end-2*tau);            % Position at time t
X2 = x(1+tau:end-tau);          % Position at time t+tau
X3 = x(1+2*tau:end);            % Position at time t+2*tau

% Plot the time series and the 3D phase-space side by side
figure;

% Plot time series of position (left side)
subplot(1,2,1);                 % Subplot (1 row, 2 columns, position
1)
plot(t, x, 'b');
xlabel('Time (t)');
ylabel('Position (x)');
title('Time Series of Harmonic Oscillator');
grid on;

% Plot the 3D phase-space (cycle plot) (right side)
subplot(1,2,2);                 % Subplot (1 row, 2 columns, position
2)
plot3(X1, X2, X3, 'r');
xlabel('x(t)');
ylabel('x(t+τ)');
zlabel('x(t+2τ)');
title('3D Phase Space of Harmonic Oscillator');
grid on;
axis equal;
view(3);                        % Set 3D view angle
```

Next we give an example of a harmonic oscillator where the amplitude gradually decreases. By executing the script 3.2 below, the results of the time series and the attractor in the reconstructed space are shown in Fig. 3.2.

```
% script 3.2
%Time series for a damped harmonic oscillator
t = 0:0.01:10;                  % Time vector (0 to 10 seconds with
0.01s intervals)
omega = 2*pi;                   % Angular frequency (1 Hz)
decay_rate = 0.1;               % Decay rate (damping factor)

% Position of the damped harmonic oscillator with exponential
decay
x = exp(-decay_rate * t) .* sin(omega * t);

% Choose the time delay (tau)
tau = 10;  % Example time delay (adjustable)

% Create the 3D phase-space (state space reconstruction)
X1 = x(1:end-2*tau);            % Position at time t
X2 = x(1+tau:end-tau);          % Position at time t+tau
X3 = x(1+2*tau:end);            % Position at time t+2*tau

% Plot the time series and the 3D phase-space side by side
figure;

% Plot time series of position (left side)
subplot(1,2,1);                 % Create a subplot (1 row, 2
columns, position 1)
plot(t, x, 'b');
xlabel('Time (t)');
ylabel('Position (x)');
title('Time Series of Damped Harmonic Oscillator');
grid on;

% Plot the 3D phase-space (cycle plot) (right side)
subplot(1,2,2);                 % Create a subplot (1 row, 2
columns, position 2)
plot3(X1, X2, X3, 'r');
xlabel('x(t)');
ylabel('x(t+τ)');
zlabel('x(t+2τ)');
title('3D Phase Space of Damped Harmonic Oscillator');
grid on;
axis equal;
view(3);                        % Set 3D view angle
```

The following is a brief presentation of the Lorenz system [11], a well-known chaotic system for specific parameter values, along with its corresponding Lorenz attractor. A more detailed reference to the Lorenz system will be made in subsequent paragraphs. By running script 3.3, we obtain the three time series describing the system evolution and its corresponding phase space representation (Fig. 3.3).

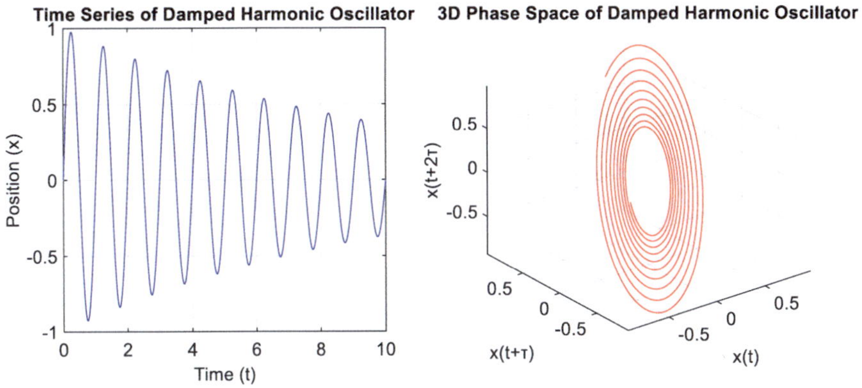

Fig. 3.2 Time series of harmonic oscillator with decreasing amplitude and the phase space attractor, respectively

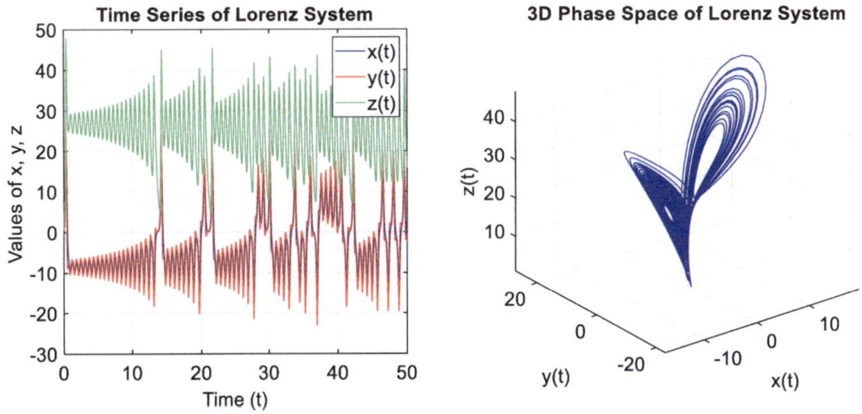

Fig. 3.3 Time series of Lorenz system and the phase space attractor, respectively

```
% script 3.3
% Lorenz system parameters
sigma = 10;
rho = 28;
beta = 8/3;

% Time span and initial conditions
tspan = [0, 50];                    % Time range for the solution
initial_conditions = [1, 1, 1];    % Initial conditions [x0, y0,
z0]

% Define the Lorenz system as a set of differential equations
lorenz = @(t, X) [ ...
    sigma * (X(2) - X(1));            % dx/dt
    X(1) * (rho - X(3)) - X(2);      % dy/dt
    X(1) * X(2) - beta * X(3)        % dz/dt
];

% Solve the system using ode45
[t, XYZ] = ode45(lorenz, tspan, initial_conditions);

% Extract the x, y, and z components
x = XYZ(:,1);
y = XYZ(:,2);
z = XYZ(:,3);

% Plot the time series and the 3D phase-space (Lorenz attractor)
side by side
figure;

% Plot the time series for x, y, and z components (left side)
subplot(1,2,1);                  % Subplot (1 row, 2 columns,
position 1)
plot(t, x, 'b', t, y, 'r', t, z, 'g');
xlabel('Time (t)');
ylabel('Values of x, y, z');
title('Time Series of Lorenz System');
legend('x(t)', 'y(t)', 'z(t)');
grid on;

% Plot the 3D phase-space (Lorenz attractor) (right side)
subplot(1,2,2);                  % Subplot (1 row, 2 columns,
position 2)
plot3(x, y, z, 'b');
xlabel('x(t)');
ylabel('y(t)');
zlabel('z(t)');
title('3D Phase Space of Lorenz System');
grid on;
axis tight;
view(3);
```

Till here, we have seen examples of phase spaces of systems where we know the governing equations. Now we will try to reconstruct the phase space based only on the knowledge of a system time series. To do so, we first need to calculate the time lag and then the estimate the embedding dimension of the system. The time lag is determined using the mutual information function, while the embedding dimension is estimated using the nearest neighbor algorithm, which is the most common method for this purpose. A widely used approach is the *False Nearest Neighbors*

(FNN) method, which identifies false neighbors based on their distance relationships. The method is described in the next paragraph.

3.1.3 False Nearest Method

The False Nearest Neighbors (FNN) is a method in phase space reconstruction used to determine the optimal embedding dimension for time series data. By embedding a time series in a higher-dimensional space, FNN identifies points that appear close in lower dimensions but are distant in higher dimensions due to projection effects. These points are labeled as "false neighbors." In this method, for an initial embedding dimension mm, we examine how the distance between two points changes when increasing the embedding dimension to $m + 1$ $m + 1$. If the distance between the points increases significantly (beyond a specified threshold), the points are considered false neighbors in the higher-dimensional space, indicating that the chosen embedding dimension mm is insufficient. This procedure is repeated for all points, and the percentage of false neighbors is analyzed to determine an appropriate embedding dimension.

The process involves incrementally increasing the embedding dimension until the fraction of false neighbors drops below a threshold, indicating that the reconstructed phase space preserves the underlying dynamics without distortions caused by insufficient dimensions. The correct dimension is identified when the percentage of false neighbors drops to near zero or below a predefined threshold, indicating that sufficient dimensions have been used to unfold the attractor. Script 3.4 [11] can be employed in Matlab to estimate the embedding dimension.

```
% script 3.4
function fnnM = FalseNearestNeighbors(xV,tauV,mV,escape,theiler)
% fnnM = FalseNearestNeighbors(xV,tauV,mV,escape,theiler)
% FALSENEARESTNEIGHBORS computes the percentage of false nearest
neighbors
% for a range of delays in 'tauV' and embedding dimensions in
'mV'.
% INPUT
%  xV        : Vector of the scalar time series
%  tauV      : A vector of the delay times.
%  mV        : A vector of the embedding dimension.
%  escape    : A factor of escaping from the neighborhood.
Default=10.
%  theiler   : the Theiler window to exclude time correlated points
in the
%              search for neighboring points. Default=0.
% OUTPUT:
%  fnnM      : A matrix of size 'ntau' x 'nm', where 'ntau' is the
number of
%              given delays and 'nm' is the number of given
embedding
%              dimensions, containing the percentage of false
nearest
%              neighbors.
```

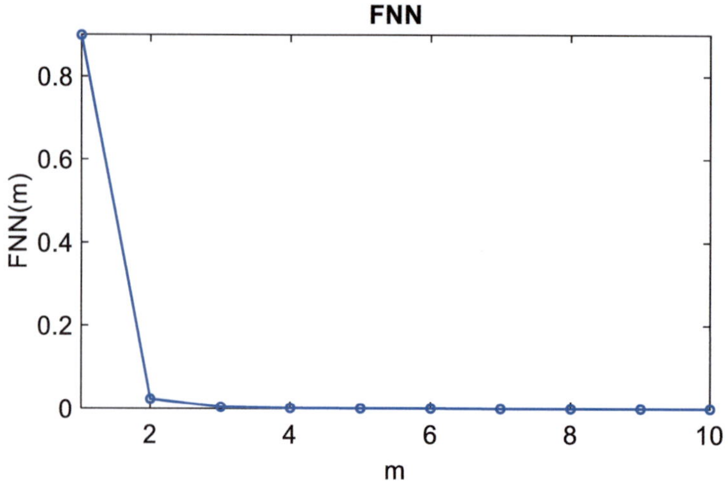

Fig. 3.4 Results of false nearest method of Lorenz time series

Using as input the time series produced for the Lorenz system previously in the txt, the embedding dimension for the case of Lorenz system can be estimated by running the following command.

```
% False Nearest Neighbors Method
>> FalseNearestNeighbors (lorenz,1,10)
```

The results appear in Fig. 3.4 where we can see that we can select $m = 3$ as embedding dimension since above this dimension the percentage of false nearest neighbors (FNN) has reached practically zero values (vertical axis).

3.1.4 Chaos and Dynamical Systems

The concept of chaos is deeply embedded in the study of complex dynamical systems, particularly within the realm of nonlinear dissipative dynamical systems [17, 18, 20]. However, while nonlinearity is a necessary condition for chaos, it is not sufficient on its own. Chaos is widely understood as the seemingly unpredictable behavior of a deterministic system due to its high sensitivity to initial conditions. This unexpected dependence on initial conditions implies that even an infinitesimal difference in initial conditions can lead to vastly different trajectories in phase space over time. As a result, long-term prediction becomes practically impossible despite the system being fully deterministic.

Examples of physical dynamical systems with chaotic behavior are found in the study of climate, astronomy, biology, atmosphere, the solar system, etc.

In chaotic dynamical systems, small variations in the initial conditions are exponentially amplified over time, leading to vastly different system evolutions. This extreme sensitivity to initial conditions, a defining characteristic of chaos, ensures that even minor differences in starting points result in trajectories that diverge exponentially in phase space. Consequently, two nearby trajectories will never converge and will continue evolving along distinct, non-repeating paths of infinite length.

Despite being governed by deterministic laws, chaotic systems are inherently unpredictable beyond a certain time horizon. This unpredictability arises because even the most precise measurements of initial conditions contain inherent limitations, making long-term forecasting practically impossible. Many natural phenomena exhibit this behavior because they originate from nonlinear dynamical systems with multiple interacting variables. In specific regions of their parameter space, these systems display chaotic dynamics, making them highly sensitive to initial conditions.

However, in practice, while chaotic systems are unpredictable in the long run, their behavior can still be estimated over short time scales before small uncertainties in initial conditions amplify beyond practical limits.

3.1.5 Dynamical Systems with an Attractor

During the temporal evolution of a dynamical system, its trajectory often remains confined within a specific region known as *basin of attraction* [3, 14]. This basin consists of all initial conditions that lead to trajectories that asymptotically converge over time toward a particular long-term behavior. Within this region, an attractor emerges as an invariant set of points to which trajectories asymptotically converge over time. The attractor's *dimension* serves as a descriptor of its geometric structure and complexity.

A special class of attractors, known as strange attractors, arises in chaotic dynamical systems, i.e., systems that exhibit sensitivity to initial conditions. In such systems, even infinitesimally close initial conditions lead to exponentially diverging trajectories, resulting in seemingly stochastic behavior despite being governed by deterministic laws. This sensitivity underpins the unpredictability of chaotic systems: they may be predictable over short time scales but become increasingly difficult to forecast as time progresses.

Strange attractors exhibit an important geometric property: self-similarity across different spatial scales, meaning they are fractal in nature. Their structure is characterized by a fractal dimension, a non-integer measure that quantifies their complexity and degree of self-similarity. The fractal dimension is always lower smaller than the topological dimension of the space in which the attractor resides. The structure of systems presenting such behavior relies on nonlinear analysis, providing insights into the behavior of chaotic systems across various scientific disciplines.

One of the most well-known dynamical systems with chaotic behavior under conditions is the Lorenz dynamical system. Lorenz created a system of three differential equations that contained two nonlinear terms and modeled heat transfer currents within a fluid. The differential equations are:

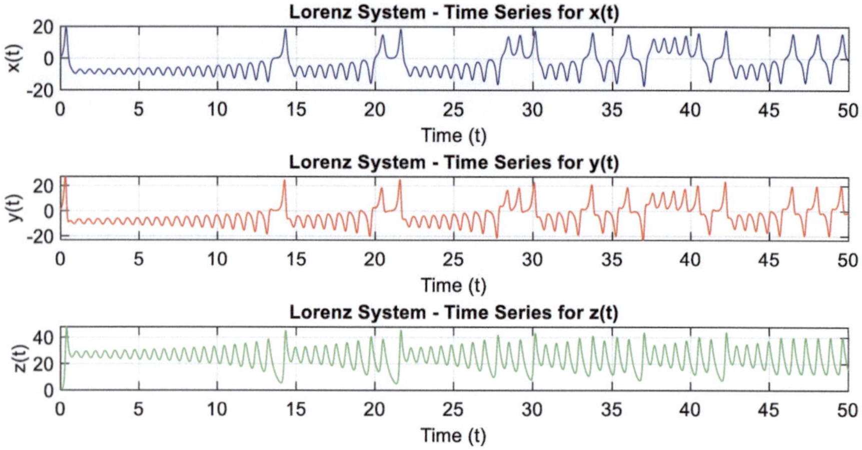

Fig. 3.5 The three time series—variables of the Lorenz system

$$\begin{aligned}\frac{dx}{dt} &= \sigma(y-x)\\ \frac{dy}{dt} &= x(\rho-z)-y\\ \frac{dz}{dt} &= xy-\beta z\end{aligned} \tag{3.4}$$

where σ, ρ, and β are three parameters of the system. More specifically, the parameter σ is called the dimensionless Prandtl number and is defined as the ratio of kinematic viscosity to thermal diffusion, the parameter ρ is called the Rayleigh number and is connected to heat transfer within a fluid and the parameter b is a geometric factor. For the values of the parameters $\sigma = 10$, $\rho = 28$, and $\beta = 8/3$, the system exhibits chaotic behavior and its numerical solution results in the creation of the trajectory in the phase space known as the Lorenz attractor. By running script 3.5, we obtain the time series that appear in Fig. 3.5 and the corresponding phase space and attractor in Fig. 3.6.

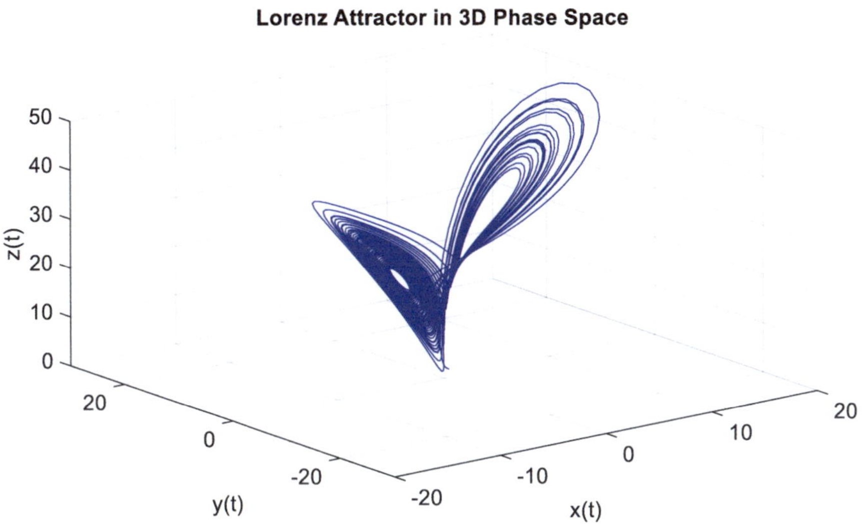

Fig. 3.6 Lorenz attractor

```
% script 3.5
% Parameters for the Lorenz system
sigma = 10;              % Parameter sigma
rho = 28;                % Parameter rho
beta = 8/3;              % Parameter beta

% Time span and initial conditions
tspan = [0, 50];                      % Time range for the solution
initial_conditions = [1, 1, 1];       % Initial values [x0, y0, z0]

% Define the Lorenz system of differential equations
lorenz_system = @(t, X) [ ...
    sigma * (X(2) - X(1));            % dx/dt
    X(1) * (rho - X(3)) - X(2);       % dy/dt
    X(1) * X(2) - beta * X(3)         % dz/dt
];

% Solve the Lorenz system using ode45
[t, XYZ] = ode45(lorenz_system, tspan, initial_conditions);

% Extract the x, y, and z components of the solution
x = XYZ(:,1);
y = XYZ(:,2);
z = XYZ(:,3);

% Plot the time series for each variable figure;

subplot(3,1,1);              % Plot for x(t)
plot(t, x, 'b');
xlabel('Time (t)');
ylabel('x(t)');
title('Lorenz System - Time Series for x(t)');
grid on;

subplot(3,1,2);              % Plot for y(t)
plot(t, y, 'r');
xlabel('Time (t)');
ylabel('y(t)');
title('Lorenz System - Time Series for y(t)');
grid on;

subplot(3,1,3);              % Plot for z(t)
plot(t, z, 'g');
xlabel('Time (t)');
ylabel('z(t)');
title('Lorenz System - Time Series for z(t)');
grid on;

% 3D Plot of the Lorenz attractor
figure;
plot3(x, y, z, 'b');
xlabel('x(t)');
ylabel('y(t)');
zlabel('z(t)');
title('Lorenz Attractor in 3D Phase Space');
grid on;
view(3);  % Set a 3D view angle for better visualization
```

3.1.6 Correlation Dimension

The calculation of the attractor's dimension is a very useful tool in nonlinear analysis and chaos theory. It helps distinguish between chaos and randomness and determines the minimum number of variables required to describe the dynamics of the system from which time series emanate [14].

An attractor, as a geometric object, is characterized by its Euclidean dimension, which described the space it occupies. In the reconstructed phase space, this dimension corresponds to the embedding dimension, denoted as m. Unlike non-chaotic attractors, which have an integer dimension equal to the topological dimension of the phase space and do not exhibit sensitivity to initial conditions, chaotic attractors are characterized by a non-integer (fractal) dimension and display the property of self-similarity across different scales of space. That fractal nature is a key signature of chaos.

The fractal dimension is expressed by the correlation dimension, which is most commonly used. Assuming two points x_i and x_j of the attractor, we define the probability $P(\|x_i - x_j\| < r)$ that their distance is less than a given radius r. If μ_i is the number of points within a sphere of radius r centered at x_i, the average over al x_i $\langle\mu_i\rangle_x$ approximates this probability. For small values of r, the scaling law applies:

$$\langle\mu_{ix}\rangle \sim r \text{ for } r \to 0$$

In the case of a time series, the average over all x_i is estimated by the correlation sum as follows:

$$C(r) = \frac{2}{N(N-1)} \sum_{i=1}^{N} \sum_{j=i+1}^{N} \Theta\left(r - \|x_i - x_j\|\right) \tag{3.5}$$

where $\Theta(x)$ is the Heaviside function. The correlation sum defines the probability that two randomly selected points of the attractor are closer than a given distance r.

For small r and using the scaling law, the correlation dimension is calculated as:

$$v = \frac{d \log C(r)}{d \log r} \tag{3.6}$$

In practice, the correlation dimension is determined as the slope of a linear region in the log-log plot of $C(r)$ versus r. Care must be taken to calculate it for a sufficient embedding dimension *mmm*. A sufficient embedding dimension corresponds to the embedding dimension m for which the correlation dimension no longer increases. Despite its widespread application, the calculated correlation dimension can be affected by the length of the time series, noise, short sampling intervals, and the choice of time delay used in reconstructing the phase space.

The correlation dimension provides information about the fractal dimension of the attractor. We evaluate the correlation dimension using the below script 3.6 [11].

```
% script 3.6 Calculation of correlation dimension

function nuT = CorrelationDimension(xV,tauV,mV,theiler,sV,resol)
% nuT = CorrelationDimension(xV,tauV,mV,theiler,sV,resol)
% CORRELATIONDIMENSION computes the correlation dimension for a
given time
% series 'xV', for a range of delays in 'tauV', a range of
embedding
% dimensions in 'mV' and for a range of upper/lower ratio of
scaling window
% in % 'sV' (s=r2/r1 where r2-r1 is the length of the scaling
window).
% The parameter 'theiler' excludes temporally close points
(smaller than
% 'theiler') from the inter-distance computations. The parameter
'resol'
% determines the number of radii for which the correlation sum is
% computed.
% First, the correlation sum C(r) and the local slopes
log(C(r))/log(r) are
% computed for a range of distances r given by 'resol'. Then the
correlation
% dimension 'nu' is estimated by searching for the local slope in
radii
% intervals [r1,r2] (determined by 's') with the smallest standard
% deviation (best scaling). The mean local slope in this interval
is the
% estimate of 'nu'.
% INPUTS:
% - xV      : Vector of the scalar time series ('xV' is then
standardized
%             in [0,1]).
% - tauV    : A vector of the delay times.
% - mV      : A vector of the embedding dimension.
% - theiler : the Theiler window to exclude time correlated points
in the
%             search for neighboring points. Default=0.
% - sV      : A vector of values of upper/lower ratio of scaling
window
%             (e=r2/r1 where r2-r1 is the length of the scaling
window).
% - resol   : The number of radius for which the correlation sum
is computed.
%             Note that this parameters is supposed to be larger
than 10.
% OUTPUT:
% - nuT     : A matrix of size 'ntau' x 'nm' x 'ne', where 'ntau'
is the
%             number of given delays, 'nm' is the number of given
embedding
%             dimensions and 'ne' is the number of scaling ratio
of radii.
%             The components of the matrix are the correlation
dimension
%             values.
```

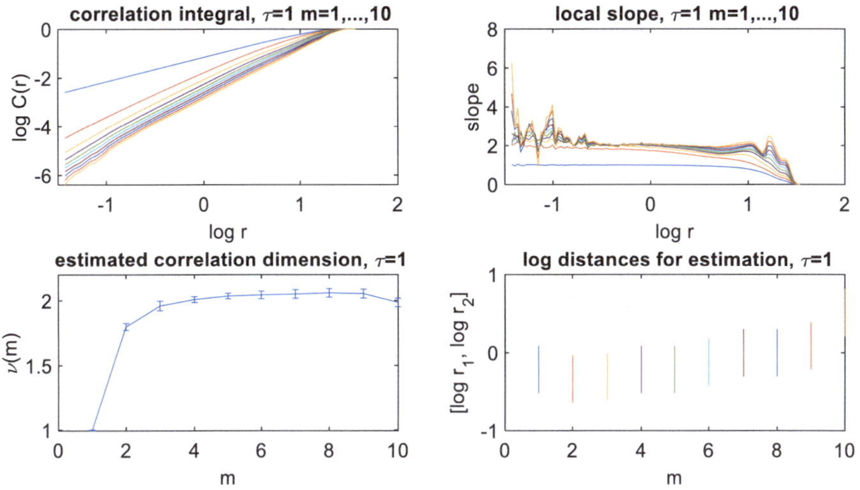

Fig. 3.7 Correlation dimension of Lorenz attractor

By running script for the time series of Lorenz attractor previously obtained, we can estimate the correlation dimension around 2.1 from Fig. 3.7. We can see in detail the log $C(r)$ vs log r diagram for which the slopes are calculated in segments and plotted and the average slope with the standard deviation for the "horizontal" parts appears as function of the embedding dimension along with the log intervals that have been employed.

3.2 Surrogate Time Series

In time series analysis, surrogate data are widely used to assess the statistical significance of results. In nonlinear time series analysis, the surrogate data method was developed to distinguish between linear stochastic processes and nonlinear deterministic dynamics [7–10, 12, 13, 19].

The core idea behind this method is to generate surrogate time series from the original data while preserving certain statistical properties. Specifically, the surrogate data method involves creating datasets that conform to a null hypothesis, allowing researchers to test whether the underlying dynamical system is linear or nonlinear.

The process begins with the hypothesis to be tested. Next surrogate time series are generated using a variety of algorithms, with the most popular ones employing the Fourier transform. A test statistic is then computed for both the original time series and the surrogates, and the statistical significance is evaluated.

An important aspect of this approach is the careful selection of the surrogate data generation algorithm since surrogate data should preserve the same cumulative distribution function and the same autocorrelation as the real data. In the following sections outline the most important techniques for generating surrogate data. To illustrate their application, each methodology is accompanied by an example demonstrating how the method works in practice.

3.2.1 Random Phase or Fourier Transform

The fundamental concept behind surrogate data, upon which most algorithms are based, is the randomization of the phases in the Fourier transform. The algorithm assumes that the time series originates from a stochastic Gaussian process. Surrogate data are constructed in such a way that they retain the same periodogram (Fourier spectrum). Initially, the Fourier transform of the time series Xt is computed for all frequencies. Then, the phases are randomized by multiplying each complex amplitude by $e^{i\Phi}$, where Φ is independently chosen for each frequency from the interval $[0, 2\pi]$. Subsequently, the inverse Fourier transform produces the final surrogate data Yt. It is important to note that, for the inverse Fourier transform to contain real components, the phases must be symmetric, such that $\Phi(f) = -\Phi(-f)$.

```
% script 3.7
% Generate surrogate time series (Random Phase or Fourier
% Transform)

% Original time series (example: a sine wave with noise)
N = 1000; % Length of the time series
t = linspace(0, 10, N); % Time vector
original_series = sin(2 * pi * 1 * t) + 0.2 * randn(1, N);

% Parameters for surrogate generation
num_surrogates = 10; % Number of surrogate series to generate
surrogate_series = zeros(num_surrogates, N);

% Generate surrogates using the Random Phase Fourier Transform
method
for s = 1:FGM_surrogates
    % Step 1: Fourier Transform of the original series
    fft_original = fft(original_series);
    magnitude = abs(fft_original);
    phase = angle(fft_original);

    % Step 2: Add random phase to the original phase
    random_phase = 2 * pi * rand(1, N) - pi; % Uniform random
phase in [-pi, pi]
    new_phase = phase + random_phase;

    % Step 3: Construct surrogate in Fourier domain
    surrogate_fft = magnitude .* exp(1i * new_phase);

    % Step 4: Inverse FFT to get the surrogate time series
    surrogate_series(s, :) = real(ifft(surrogate_fft));
end

% Plot original series and surrogates
figure;
subplot(2, 1, 1);
plot(t, original_series, 'b', 'LineWidth', 1.5);
title('Original Time Series');
xlabel('Time'); ylabel('Amplitude'); grid on;

subplot(2, 1, 2);
hold on;
for s = 1:num_surrogates
    plot(t, surrogate_series(s, :), 'LineWidth', 1);
end
title('Surrogate Time Series');
xlabel('Time'); ylabel('Amplitude'); grid on;
hold off;
```

Using the above script (3.7), we construct ten surrogate time series with the random phase Fourier transform method. In Fig. 3.8, we can see the original time series and the corresponding surrogate time series.

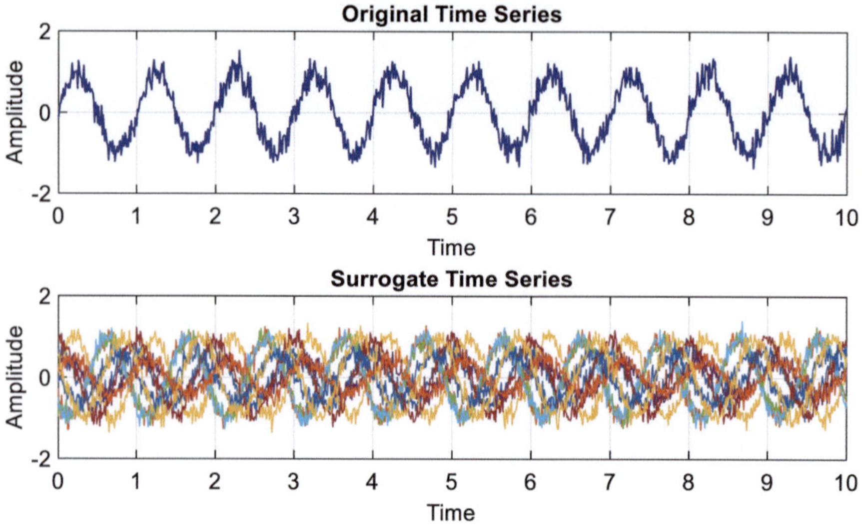

Fig. 3.8 Original time series and the surrogates using random phase Fourier transform method

3.2.2 Amplitude Adjusted Fourier Transform (AAFT)

The Amplitude Adjusted Fourier Transform (AAFT) generates surrogate data by using a random time series with a normal distribution. The elements of the time series are then rearranged to match the rank ordering of the original time series (Xt) (rank ordering white noise), thus producing the time series Xt′. Next, the Fourier transform of Xt′ is calculated, and the phases are randomly shuffled to ensure that the surrogate time series maintains the same power spectrum as the original. This process results in a new time series Xt″. Subsequently, the inverse Fourier transform of Xt″ is computed. By observing the rank order distribution of the time series elements, the data from the original time series are rearranged based on the data of Xt″.

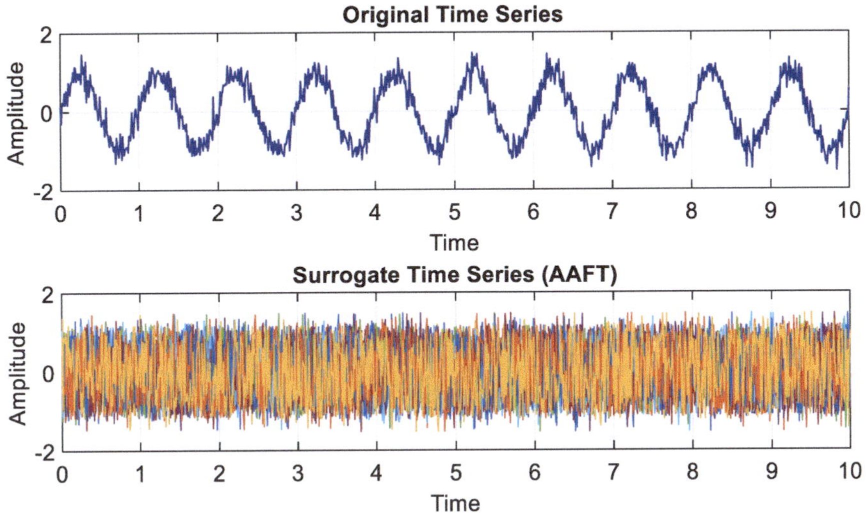

Fig. 3.9 Original time series and the surrogates using amplitude adjusted Fourier transform method

In this way, the surrogate time series have the same distribution and approximately the same power spectrum. However, because the algorithm assumes that the static transformation is monotonic, it fails to accurately reproduce the linear correlations (autocorrelation function) in the surrogate time series. This means that if the original time series does not employ a monotonic static transformation, there may be discrepancies in the linear correlations between the surrogate and the original time series. As a result, the AAFT algorithm is used to test the null hypothesis that the time series originates from a Gaussian process subjected to a monotonic static transformation.

By running in Matlab command window, the following script (3.8) we get the results appearing in Fig. 3.9, where the original time series and the surrogates are presented.

```
% script 3.8
% Generate surrogate time series (Amplitude Adjusted Fourier
%  Transform (AAFT))

% Original time series (example: a sine wave with noise)
N = 1000; % Length of the time series
t = linspace(0, 10, N); % Time vector
original_series = sin(2 * pi * 1 * t) + 0.2 * randn(1, N);

% Parameters for surrogate generation
num_surrogates = 10; % Number of surrogate series to generate
surrogate_series = zeros(num_surrogates, N);

% Generate surrogates using the Amplitude Adjusted Fourier
Transform (AAFT) method
for s = 1:num_surrogates
    % Step 1: Rank-order original series to create a Gaussian-
distributed series
    sorted_original = sort(original_series);
    gaussian_series = sort(randn(1, N));
    [~, idx] = sort(original_series);
    rank_ordered_series = gaussian_series(idx);

    % Step 2: Fourier Transform of the rank-ordered series
    fft_gaussian = fft(rank_ordered_series);
    magnitude = abs(fft_gaussian);
    phase = angle(fft_gaussian);

    % Step 3: Add random phase to preserve spectrum
    random_phase = 2 * pi * rand(1, N) - pi; % Uniform random
phase in [-pi, pi]
    surrogate_fft = magnitude .* exp(1i * random_phase);

    % Step 4: Inverse FFT to get the surrogate series
    gaussian_surrogate = real(ifft(surrogate_fft));

    % Step 5: Match the amplitude distribution of the original
series
    [~, surrogate_idx] = sort(gaussian_surrogate);
    amplitude_adjusted_series = zeros(1, N);
    amplitude_adjusted_series(surrogate_idx) = sorted_original;

    % Store the surrogate series
    surrogate_series(s, :) = amplitude_adjusted_series;
end

% Plot original series and surrogates
figure;
subplot(2, 1, 1);
plot(t, original_series, 'b', 'LineWidth', 1.5);
title('Original Time Series');
xlabel('Time'); ylabel('Amplitude'); grid on;

subplot(2, 1, 2);
hold on;
for s = 1:num_surrogates
    plot(t, surrogate_series(s, :), 'LineWidth', 1);
end
title('Surrogate Time Series (AAFT)');
xlabel('Time'); ylabel('Amplitude'); grid on;
hold off;
```

```
% script 3.9
% Generate surrogate time series (Iterative Amplitude Adjusted
Fourier Transform (IAAFT))

% Original time series (example: a sine wave with noise)
N = 1000; % Length of the time series
t = linspace(0, 10, N); % Time vector
original_series = sin(2 * pi * 1 * t) + 0.2 * randn(1, N);

% Parameters for surrogate generation
num_surrogates = 10; % Number of surrogate series to generate
surrogate_series = zeros(num_surrogates, N);

% Generate surrogates using the Iterative Amplitude Adjusted
Fourier Transform (IAAFT) method
for s = 1:FGM_surrogates
    % Step 1: Rank-order original series to create a Gaussian-
distributed series
    sorted_original = sort(original_series);
    gaussian_series = sort(randn(1, N));
    [~, idx] = sort(original_series);
    rank_ordered_series = gaussian_series(idx);

    % Step 2: Initialize with Fourier Transform of the rank-
ordered series
    fft_gaussian = fft(rank_ordered_series);
    magnitude = abs(fft_gaussian);
    phase = angle(fft_gaussian);

    % Iterative adjustment
    max_iter = 100;
    tol = 1e-6;
    surrogate = rank_ordered_series;
    for iter = 1:max_iter
        % Fourier Transform of the surrogate series
        fft_surrogate = fft(surrogate);

        % Enforce original magnitude spectrum
        adjusted_fft = magnitude .* exp(1i *
angle(fft_surrogate));
        adjusted_series = real(ifft(adjusted_fft));

        % Rescale to match the amplitude distribution of the
original series
        [~, surrogate_idx] = sort(adjusted_series);
        rescaled_series = zeros(1, N);
        rescaled_series(surrogate_idx) = sorted_original;

        % Check for convergence
        if max(abs(rescaled_series - surrogate)) < tol
            break;
        end
        surrogate = rescaled_series;
    end

    % Store the final surrogate series
    surrogate_series(s, :) = surrogate;
end
```

3.2.3 *Iterative Amplitude Adjusted Fourier Transform (IAAFT)*

The Iterative Amplitude Adjusted Fourier Transform (IAAFT) algorithm improves upon the AAFT algorithm by more accurately preserving the linear correlations of surrogate time series through an iterative refinement process.

The procedure begins by randomly shuffling the values of the original time series. Then, a surrogate time series is generated to match the original **power spectrum** while preserving the rank order of the data. This is achieved by generating a white noise time series and replacing its squared magnitudes with those of the original time series.

Subsequently, since the power spectrum is preserved but the marginal distribution is altered, the elements are reordered to match the rank order of the original time series. However, since this process achieves marginal distribution but modifies the power spectrum, the previous steps are repeated iteratively until convergence is achieved in both the correlations of the data and the marginal distribution. Thus, the surrogate time series have almost the same distribution and linear structure as the original time series. However, due to the algorithm's mechanism, the surrogate time series are consistently less correlated than the original time series, which may result in significant linear discrepancies.

In general, the IAAFT algorithm is used to test the null hypothesis that the time series originates from a Gaussian process subjected to a static transformation. Using the script (3.9), we generate the surrogates time series presented in Fig. 3.10.

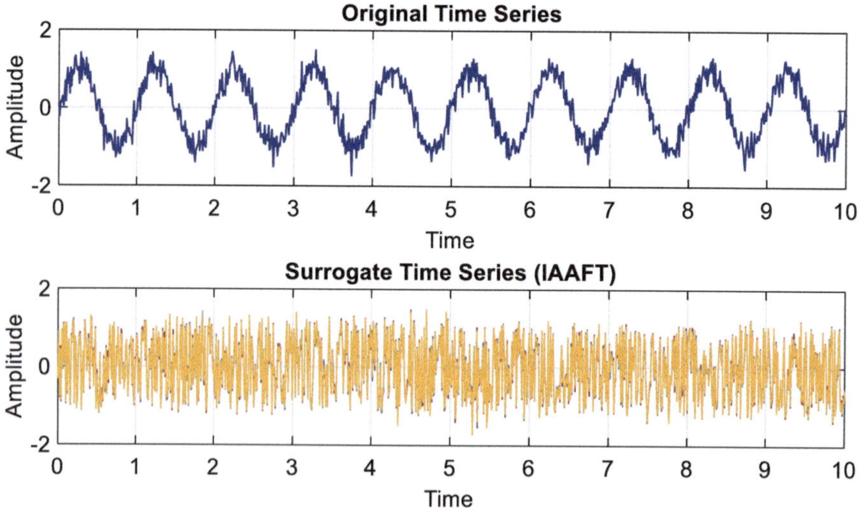

Fig. 3.10 Original time series and the surrogates using iterative amplitude adjusted Fourier transform (IAAFT) method

```
% Plot original series and surrogates
figure;
subplot(2, 1, 1);
plot(t, original_series, 'b', 'LineWidth', 1.5);
title('Original Time Series');
xlabel('Time'); ylabel('Amplitude'); grid on;

subplot(2, 1, 2);
hold on;
for s = 1:num_surrogates
    plot(t, surrogate_series(s, :), 'LineWidth', 1);
end
title('Surrogate Time Series (IAAFT)');
xlabel('Time'); ylabel('Amplitude'); grid on;
hold off;
```

3.2.4 Statistically Transformed Autoregressive Process (STAP)

The Statistically Transformed Autoregressive Process (STAP) algorithm generates surrogate data that preserve both the autocorrelation structure and the probability distribution function of the original time series. In a first place a white Gaussian noise time series is created, and the transformation from the Gaussian to the desired distribution is approximated using a polynomial of a specified degree (m). The autocorrelation of an autoregressive process u (AR(p)) is represented as a polynomial function of degree mm of the autocorrelation of the original time series.

The parameters of the AR model of order p are estimated using the Levinson algorithm, leading to the construction of the u time series based on the AR(p) model. This series is then transformed into the surrogate data time series, ensuring it shares the same linear structure (autocorrelation) and probability distribution function as the original time series. The procedure is repeated to optimize the autocorrelation match with the original time series.

Thus, surrogate time series generated by the STAP algorithm exhibit greater variability than those produced by the IAAFT algorithm, while maintaining the same linear structure as the original time series. In general, the STAP algorithm is used to test the null hypothesis that the time series originates from a Gaussian process that has undergone a static transformation. Figure 3.11 presents the surrogates time series produced using script 3.10.

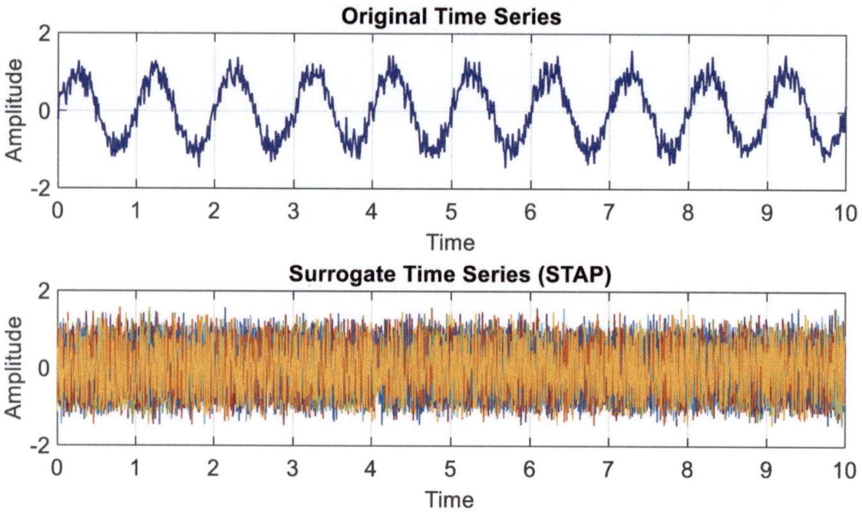

Fig. 3.11 Original time series and the surrogates using statistically transformed autoregressive process (STAP) method

```
% script 3.10
% Generate surrogate time series (Statistically Transformed
% Autoregressive Process (STAP))

% Original time series (example: a sine wave with noise)
N = 1000; % Length of the time series
t = linspace(0, 10, N); % Time vector
original_series = sin(2 * pi * 1 * t) + 0.2 * randn(1, N);

% Parameters for surrogate generation
num_surrogates = 10; % Number of surrogate series to generate
surrogate_series = zeros(num_surrogates, N);

% Fit an autoregressive (AR) model to the original series
order = 2; % Order of the AR model
ar_model = ar(original_series, order); % Estimate AR model
coefficients

% Generate surrogates using Statistically Transformed
Autoregressive Process (STAP)
for s = 1:num surrogates
    % Step 1: Generate a Gaussian white noise sequence
    gaussian_noise = randn(1, N);

    % Step 2: Use the AR model to generate an AR process
    ar_process = filter(1, [1 -ar_model.a(2:end)],
gaussian_noise);

    % Step 3: Match the amplitude distribution of the original
    % series
    [sorted_original, idx_original] = sort(original_series);
    [sorted_ar, idx_ar] = sort(ar_process);
    transformed_process = zeros(1, N);
    transformed_process(idx_ar) = sorted_original;

    % Store the surrogate series
    surrogate_series(s, :) = transformed_process;
end

% Plot original series and surrogates
figure;
subplot(2, 1, 1);
plot(t, original_series, 'b', 'LineWidth', 1.5);
title('Original Time Series');
xlabel('Time'); ylabel('Amplitude'); grid on;

subplot(2, 1, 2);
hold on;
for s = 1:num_surrogates
    plot(t, surrogate_series(s, :), 'LineWidth', 1);
end
title('Surrogate Time Series (STAP)');
xlabel('Time'); ylabel('Amplitude'); grid on;
hold off;
```

References

1. Brockwell, P. J. (1991). *Time series: Theory and methods*. Springer-Verlag.
2. Charakopoulos, A. K., Karakasidis, T. E., Papanicolaou, P. N., & Liakopoulos, A. (2014). Nonlinear time series analysis and clustering for jet axis identification in vertical turbulent heated jets. *Physical Review E, 89*(3), 032913.
3. Elsner, J. B., & Tsonis, A. A. (1996). Phase space reconstruction. In *Singular spectrum analysis* (pp. 143–155). Springer US.
4. Fraser, A. M., & Swinney, H. L. (1986). Independent coordinates for strange attractors from mutual information. *Physical Review A, 33*(2), 1134.
5. Grassberger, P., & Procaccia, I. (1983). Measuring the strangeness of strange attractors. *Physica D: Nonlinear Phenomena, 9*(1–2), 189–208.
6. Kantz, H., & Schreiber, T. (2003). *Nonlinear time series analysis*. Cambridge University Press.
7. Kugiumtzis, D. (1999). Test your surrogate data before you test for nonlinearity. *Physical Review E, 60*(3), 2808–2816.
8. Kugiumtzis, D. (2000). Surrogate data test for nonlinearity including nonmonotonic transforms. *Physical Review E, 62*(1), R25.
9. Kugiumtzis, D. (2008). Evaluation of surrogate and bootstrap tests for nonlinearity in time series. *Studies in Nonlinear Dynamics & Econometrics, 12*(1).
10. Kugiumtzis, D., & Tsimpiris, A. (2010). Measures of analysis of time series (MATS): A MATLAB toolkit for computation of multiple measures on time series data bases. *Journal of Statistical Software, 33*, 1–30.
11. Lorenz, H. W. (1993). *Nonlinear dynamical economics and chaotic motion* (Vol. 334). Springer.
12. Schreiber, T., & Schmitz, A. (1996). Improved surrogate data for nonlinearity tests. *Physical Review Letters, 77*(4), 635.
13. Schreiber, T., & Schmitz, A. (2000). Surrogate time series. *Physica D: Nonlinear Phenomena, 142*(3–4), 346–382.
14. Sivakumar, B. (2004). Chaos theory in geophysics: Past, present and future. *Chaos, Solitons & Fractals, 19*(2), 441–462.
15. Sprott, J. (2003). *Chaos and time-series analysis*. Oxford University Press.
16. Strogatz, S. H. (2014). *Nonlinear dynamics and chaos: With applications to physics, biology, chemistry, and engineering*. Westview press.
17. Takens, F. (2006, October). Detecting strange attractors in turbulence. In *Dynamical systems and turbulence, Warwick 1980: Proceedings of a symposium held at the University of Warwick 1979/80* (pp. 366–381). Springer Berlin Heidelberg.
18. Theiler, J. (1990). Estimating fractal dimension. *JOSA A, 7*(6), 1055–1073.
19. Theiler, J., Eubank, S., Longtin, A., Galdrikian, B., & Farmer, J. D. (1992). Testing for nonlinearity in time series: The method of surrogate data. *Physica D: Nonlinear Phenomena, 58*(1–4), 77–94.
20. Tsonis, A. A. (2012). *Chaos: From theory to applications*. Springer Science & Business Media.

Part II
Complex Network Analysis

Chapter 4
Complex Network Time Series

Complex network-based time series analysis has gained significant attention, and they have provided new insights to time series analysis by applying graph theory. This approach has achieved valuable results in addressing interdisciplinary challenges. Network analysis findings indicate that time series properties can be inherited by the network measures, enhancing our understanding of crucial processes in physics, economy, among others such as neuroscience, biology, medicine, and finance [1, 3, 4, 5, 8].

4.1 Basics of Complex Network Theory

In this part of the book, we present some of the basic ideas behind graph theory and the key aspects of the study of network structure. Network analysis is based on a branch of mathematics called graph theory to define the basics concepts.

4.1.1 Theoretical Definition of a Graph

A graph consists of a set of objects, called nodes or vertices, with certain pairs of these objects connected by links called edges. Mathematically, a graph $G = (N;E)$ is defined by a set of nodes N and a set of edges E connecting them [1, 3].

A vertex (singular of vertices) is the smallest unit in a network and is typically labeled with a number, while an edge represents a connection (or tie) between two vertices.

For example, the graph in Fig. 4.1 consists of nine nodes labeled 1, 2, 3, …, 9. Two nodes are considered neighbors if they are connected by an edge. Figure 4.1

T. Karakasidis, A. Charakopoulos, *Time Series and Networks Analysis*,
https://doi.org/10.1007/978-3-031-92628-0_4

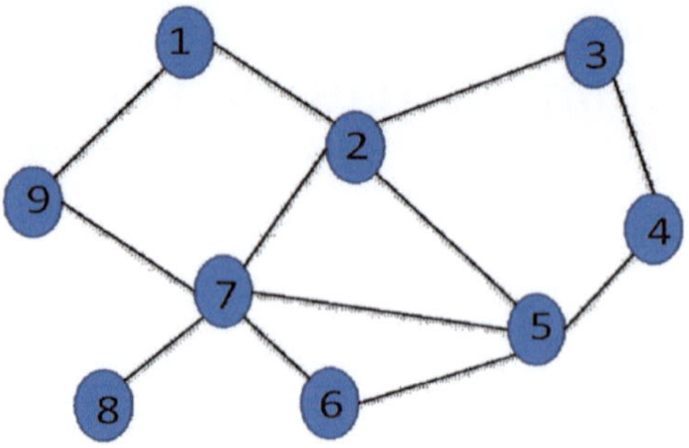

Fig. 4.1 Network consisting of nine nodes

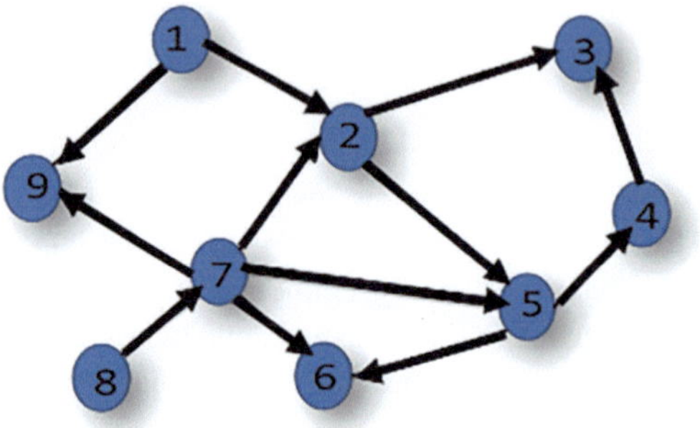

Fig. 4.2 Directed network consisting of nine nodes

illustrates the typical way one draws a graph with small circles representing the nodes, and lines connecting them indicate edges.

Networks can be classified into two basic types based on node connectivity: directed networks and undirected networks. A directed graph is defined as a type of graph where edges have an assigned direction. In contrast, an undirected graph has edges with no specified direction. In directed graphs, edges are called arcs, which are ordered pairs of vertices where the first vertex is the sender and the second is the receiver. In undirected graphs, edges are represented as unordered pairs, meaning the order of vertices does not matter. A directed graph contains one or more arcs, whereas an undirected graph contains no arcs, only bidirectional edges. Figure 4.1 illustrates an undirected graph, while Fig. 4.2 shows a directed network.

From a mathematical point of view, we can represent a network using an adjacency matrix A. A graph with N vertices can be represented by an $N \times N$ adjacency matrix, where each entry indicates the presence or absence of an edge between two vertices. Figure 4.3 presents an undirected network and its corresponding adjacency matrix.

Another distinction that may occur between graphs are the unweighted and weighted graphs. An unweighted graph is a graph in which all edges are considered equal, meaning they do not carry any numerical values or weights. The presence or absence of an edge simply indicates whether two vertices are connected. In an unweighted adjacency matrix, entries are typically 1 (if an edge exists) or 0 (if no edge exists). In contrast, a weighted graph is a graph in which each edge is assigned

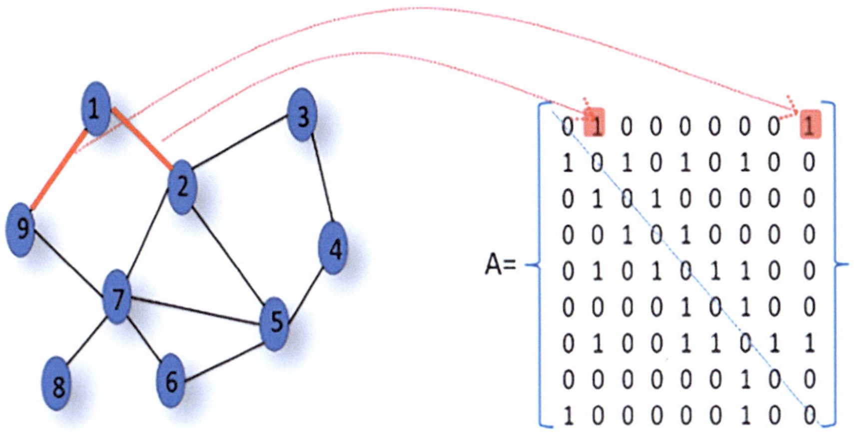

Fig. 4.3 Undirected network and the corresponding adjacency matrix

a numerical value, called a weight, which represents some measure of the connection between two vertices. Weights can represent distances, costs, capacities, probabilities, or other relevant quantities depending on the context. In a weighted adjacency matrix, entries store the weight of each edge instead of just 1s and 0s.

For an undirected and unweighted graph, like the one shown in Fig. 4.3, the edges can be represented by the elements A_{ij} of this matrix such $A_{ij} = 1$, if the nodes i and j are connected, otherwise it is 0. Two nodes joined by a link are referred to as adjacent or neighboring nodes. In this case, the adjacency matrix is symmetric, it means $A_{ij} = A_{ji}$, while for directed ones, the matrix is not symmetric and the element $A_{ij} = 1$ indicates that the node i points to the node j.

An example of adjacency matrix in the Matlab can be obtained by executing the script 4.1 in the command window below, and the result is displayed in Fig. 4.4.

```
% Script 4.1
% Suppose we have a network name  A_net1

Inp1=input('name of adjacency_');

spy(Inp1)

title('Adjacency matrix','FontSize',20)
ylabel('nodes')
xlabel('nodes')
```

We can see that the adjacency matrix is symmetrical with respect to the main diagonal (the dots correspond to the number 1, while the spaces correspond to the number 0). As we have mentioned, the existence of a dot indicates that there is a connection between the nodes. So, node 1 is connected with node 2, 4, 5, 6, and 7.

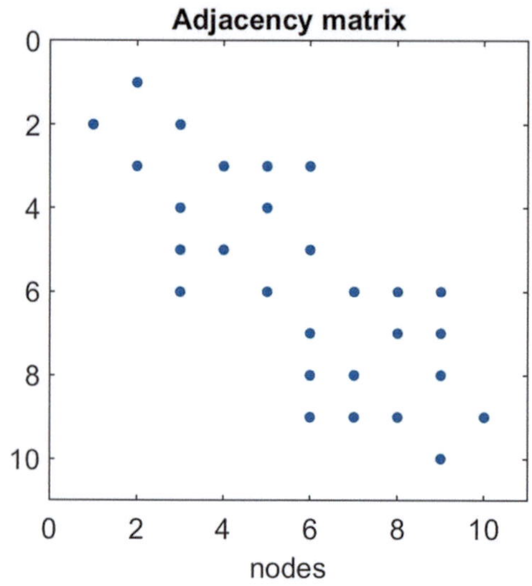

Fig. 4.4 Adjacency matrix of a network consisting of ten nodes

4.2 Topological Network Measures

As previously mentioned, an undirected and unweighted complex network can be represented as a graph $G = (N,E)$ consisting of a set of $N = (n_1,n_2,...n_N)$ nodes or vertices and a set of $E = (e_1,e_2,...,e_E)$ edges or links. The network's topological structure is described by a $N \times N$ adjacency matrix $A = [a_{ij}]$ where $a_{ij} = 1$ if the vertex i is connected to vertex j and $a_{ij} = 0$ otherwise.

4.2.1 Degree and Degree Distribution

Degree of a Node

The degree of a node i (k_i) is a fundamental parameter of a network that influences other characteristics and is defined as the total number of connections (or edges) adjacent to that node i. Hence, the degree of a node is the number of edges that it shares with other nodes [1, 3, 5].

For undirected networks, it can be computed as

$$k_i = \sum_j a_{ij} = \sum_j a_{ji} \tag{4.1}$$

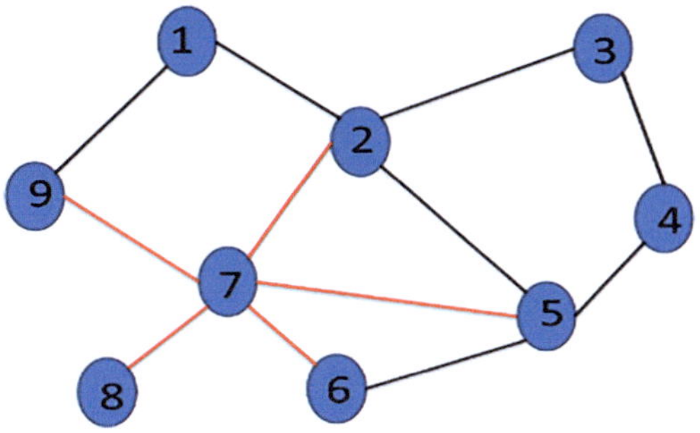

Fig. 4.5 Example of degree of a network node

characterizing the connectivity properties of an individual node in a network, the average degree $\langle k \rangle$ represents the mean value of k_i across all nodes. It serves as a global measurement of the connectivity of the network.

$$\langle k \rangle = \frac{1}{N}\sum_i k_i = \frac{1}{N}\sum_{ij} a_{ji} \tag{4.2}$$

The average degree of all vertices is a measure of the structural cohesion of a network. In the case of a directed network, we can define the in-degree and the out-degree. The in-degree of a node is the number of arcs it receives, while the out-degree is the number of arcs it sends.

Figure 4.5 illustrates an example of node calculation in the case of an undirected network. The degree of node 7 is $k_7 = 5$ since node 7 is connected with five nodes, i.e., N_2, N_5, N_6, N_8, and N_9. The degree of node 2 is $k_2 = 4$ since it is connected to four nodes, i.e., N_1, N_7, N_3, N_5.

Let's assume that we have a network consisting of 100 nodes (Fig. 4.6), for which we calculate the degree of each node as well as the average degree overall for the network.

Using the following script (4.2) in Matlab, we can calculate the degree of each node in the network.

```
% Script 4.2
% Example calculation of node degree

function [Deg] = Degree_network(adj)

%   Node degree is the number of links connected to the node.

%   Input:      adj,    adjacency matrix

%   Output:     Deg,    node degree

adj = double(adj~=0);

 Deg = sum(adj);
 figure;
 stem(Deg)
 xlabel('node');
 ylabel('degree of node');
 title('Value of Node Degree of a Network');
```

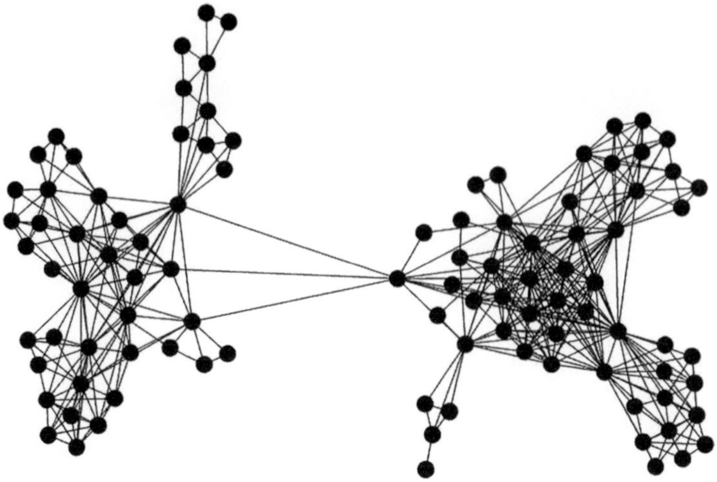

Fig. 4.6 Example of a network consisting of 100 nodes

In the Matlab command, we have to type the following:

```
>>Degree_network(Net_A1) %  Net_A1 is the name of adjacency matrix
```

or we can execute the above script.

```
% Suppose we have a network name Net_A1

Inp1=input('name of adjacency_');

Degree_network(Inp1)
```

Figure 4.7 shows the results of the node degree (number of connections) of each node for the network of Fig. 4.6. The height of each bar indicates the degree of each node, that is, the number of connections of each node. From this graph, we can see which node has the most connections. We can see that node 12 has the highest number of connections and thus the highest degree (with value equal to 31).

Degree Distribution

The **degree distribution** of a network is the fraction of nodes with degree *kkk*, representing the probability that a randomly selected node has exactly *kkk* edges. Examining the degree distribution—often visualized as a histogram—provides insight into the structure and behavior of the network. Another definition is that

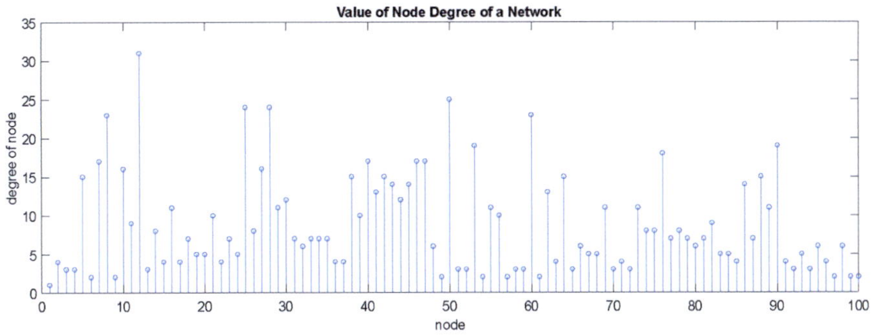

Fig. 4.7 Degree of network consisting of 100 nodes

degree distribution is the frequency distribution of different degrees across the nodes in the network and can be used to characterize a network. The degree distribution gives the probability that a selected node has exactly *k* edges.

Below the code (script 4.3) for calculating the degree distribution is given.

```
% Script 4.3
% Example of node degree and degree distribution

function [DED,Aver_DED]=Degree_Distribution_net(adj)

N=size(adj,2);
DED=zeros(1,N);
for i=1:N
   DED(i)=sum(adj(i,:));
end
Aver_DED=mean(DED);

if sum(DED)==0
    disp('............');
    return;
else
    figure;
    subplot(1,2,1);
    bar([1:N],DED);
    xlabel('Node');
    ylabel('degree of node');
    title('Node Degree of a Network');
    grid on;
    subplot(1,2,2);
    M=max(DED);
for i=1:M+1;
    N_DeD(i)=length(find(DED==i-1));
end
P_DeD=zeros(1,M+1);
P_DeD(:)=N_DeD(:)./sum(N_DeD);
loglog([0:M],P_DeD);
xlabel('node degree k'); ylabel('Number of nodes');
title('Degree Distribution');
grid on;
end
```

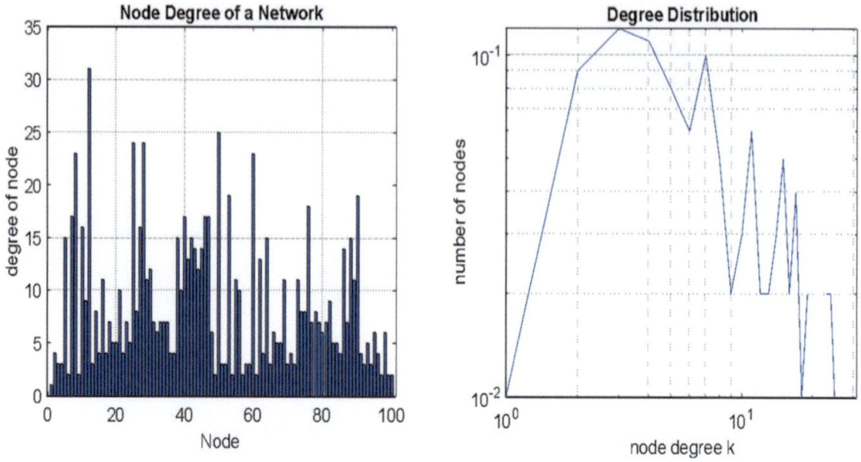

Fig. 4.8 Degree per node (left) and degree distribution (right) of a network consisting of 100 nodes

In the Matlab command, we write:

```
>> Degree_Distribution_net(Net_A1)
%  Net_A1 is the name of adjacency matrix
```

or we can execute the script file (4.3) and the results appear in Fig. 4.8. Pay attention that the horizontal axis on the degree distribution (right figure) is logarithmic. We calculate the degree of each node, i.e., the connections, as well as the degree distribution of points.

4.2.2 *Shortest Path and Diameter*

In a network, the number of edges in a path connecting vertices *i* and *j* is called the length of the path. The distance d_{ij} between nodes *i*, *j* defined as the length of the shortest path connecting them. The network's diameter *D* is the longest shortest path distances between any pair of nodes of a network. The average distance $<d_{i,j}>$ is the average distance of a network connecting any pair of points *i*, *j*. In other words, the average path length is the mean number of edges in the shortest paths connecting all nodes in the network:

$$D = \max_{i,j} d_{i,j} \tag{4.3}$$

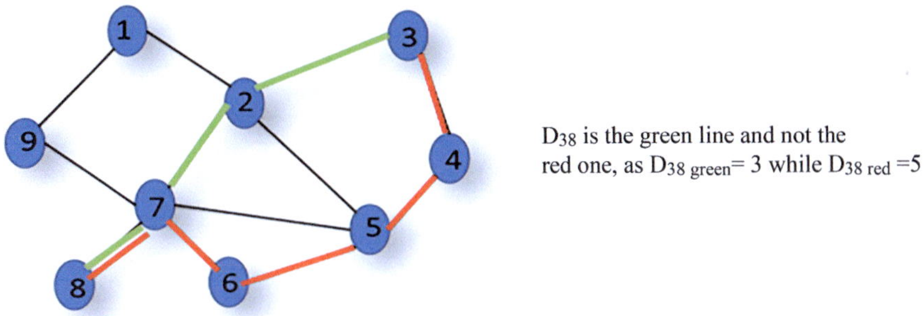

Fig. 4.9 Example of distance in a network

$$\langle d_{i,j} \rangle = \frac{1}{N(N-1)} \sum_{i,j} d_{i,j} \tag{4.4}$$

Diameter is an important characteristic that depends on the overall network structure.

In Fig. 4.9, we present a network with examples of distances.

To calculate the network diameter of the network, the following script needs to be executed in Matlab.

```
% Script 4.4
% Example of network diameter / The longest shortest path between
any two nodes nodes in the network

% INPUTS: adjacency matrix, adj

% OUTPUTS: network diameter, diam

function [diam,dij] = Diameter_Net(adj)

diam=0;
dij=[];
for i=1:size(adj,1)
    d=simple_dijkstra(adj,i);
    dij=[dij;d];
    diam = max([max(d),diam]);
end
```

In the command window of Matlab, we type the following command:

```
>> Diameter_Net(Net_A1) %  Net_A1 is the name of adjacency matrix

ans= 6                    % ans the result of diameter of network
```

Figure 4.10 shows the shortest path between nodes in the network of Fig. 4.6. Specifically, we can see two shorter paths: the first where it connects the two distant red nodes and the one that connects the two blue nodes.

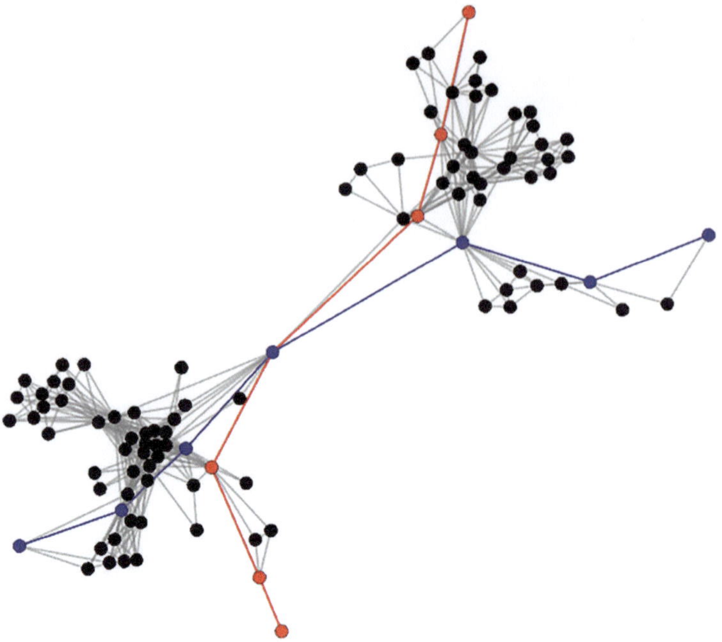

Fig. 4.10 Example of shortest paths of network consisting of 100 nodes

4.2.3 Clustering Coefficient

The clustering coefficient of a network quantifies the local link density by counting the triangles in the network and for a node i can be defined as:

$$c_i = \frac{2e_i}{k_i\left(k_i - 1\right)} \tag{4.5}$$

where k_i is the number of neighbors of i and e_i is the number of connected pairs between all neighbors of i in a network. We can say that if a node i has k_i neighbors or friends, then all possible connections among neighbors pairs of the nodes in a graph are $k_i(k_i - 1)/2$.

The average clustering coefficient C of a network is the average c_i and it is defined as

$$C = \langle c_i \rangle = \frac{1}{N}\sum_i c_i \tag{4.6}$$

An alternative definition of the clustering coefficient of a node i is the ratio E/M, where E is the number of edges between the neighbors of node i, and M is the maximum number of edges that could potentially exist between those neighbors. The clustering coefficient of a node ranges between 0 and 1. Therefore, the local clustering coefficient measures the probability of the neighbors of a node i being connected which is the probability that first step neighbors of a node i (called "friends" of the node i) are connected directly to each other. Clustering coefficient is a measure of network transitivity, which indicates how much neighbors of a node are neighbors of each other. Transitivity measures the probability that adjacent vertices of a node are connected. A network is considered transitive, if for any three nodes a, b, and c, when there is an edge between a and b, and between b and c, then there exists an edge between a and c as well.

Figure 4.11 illustrates clustering coefficient for a representative node. In this case, node 2 presents a degree equal to 4 and thus based on Eq. (4.5), the corresponding clustering is found equal to 1/6.

In the following, we present a Matlab script (4.5) that calculates the clustering coefficient of a network.

```
% Script 4.5
% Example of network clustering coefficient

function [Cl, avCl]=clustering_coef_network(adj)


%   The clustering coefficient is the fraction of triangles around
a node

%   Input:     adj      adjacency matrix
%
%   Output:     Cl      clustering coefficient vector per node


n=length(adj);
Cl=zeros(n,1);

for u=1:n
    V=find(adj(u,:));
    k=length(V);
    if k>=2;                  %degree must be at least 2
        S=adj(V,V);
        Cl(u)=sum(S(:))/(k^2-k);
    end

end
avCl=mean(Cl) %average clustering coefficient

figure;
 stem(Cl)
 ylim([0 1.5])
 xlabel('node');
 ylabel('clustering coefficient of node');
 title('Value of Clusteirng Coefficient of a Network node');
```

In the command window of Matlab, we type the following commands and the results are presented in Fig. 4.12, where we can see the value of the clustering

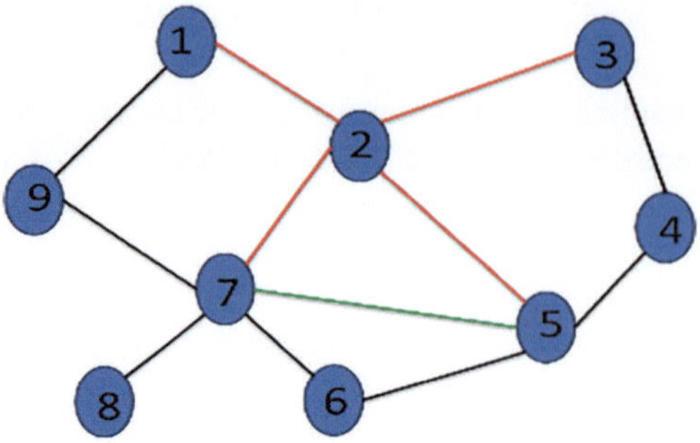

$$C_{(Vi)} = \frac{2 \cdot n}{k_i \cdot (k_i - 1)}$$

$$C_{(V2)} = \frac{2 \cdot 1}{4 \cdot (4-1)} = \frac{1}{6}$$

Fig. 4.11 Example of clustering coefficient of a network

```
>> [Cl]=clustering_coef_network(Net_A1)

% Net_A1 is the name of adjacency matrix

avCl =

    0.7467

% avCl is the average Clustering Coefficient of the network
```

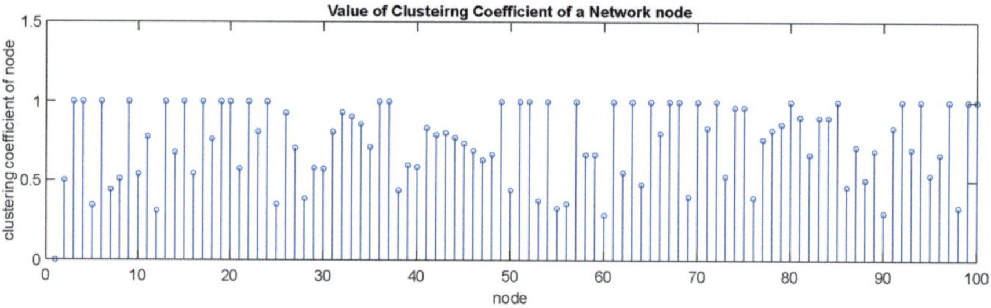

Fig. 4.12 Clustering coefficient of the nodes of a network consisting of 100 nodes

coefficient for each node of the network. In Fig. 4.13, the network nodes are represented in a color scale based on their value of clustering coefficient. The darker the node, the higher the value of the clustering coefficient.

```
>> [Cl]=clustering_coef_network(Net_A1)

% Net_A1 is the name of adjacency matrix

avCl =

    0.7467

% avCl is the average Clustering Coefficient of the network
```

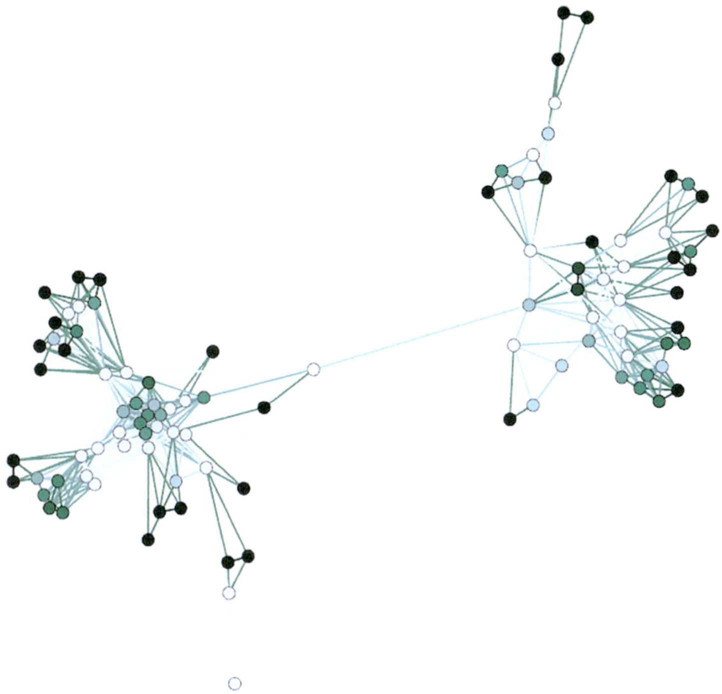

Fig. 4.13 Graphical representation of clustering coefficients of nodes of network consisting of 100 nodes

4.2.4 *Centrality Measures/Betweenness Centrality*

Centrality Measures

Centrality measures indicating the importance of node in the network (is deciding on whether there are any vertices "more important" than others). Below we discuss the various measures.

Betweenness Centrality

Betweenness centrality is calculated based on the position of the node in the network paths. Thus, nodes with high betweenness can have significant influence within a network for the transmission of information. The betweenness centrality of a node v is given by the following equation:

$$g_{(V)} = \sum_{s \neq v \neq t} \frac{\sigma_{st}(v)}{\sigma_{st}} \tag{4.7}$$

where σ_{st} is the total number of shortest paths from node s to node t and $\sigma_{st}(v)$ is the number of those paths that pass-through v.

The following script (4.6) can be used in Matlab to calculate the betweenness centrality.

```
% Script 4.6
% Example of network Betweenness Centrality

% Betweenness centrality measure: Number of shortest paths
running though a vertex

% INPUTS: adjacency  matrix

% OUTPUTS: betweeness vector for all vertices


function betw = betweenness_centrality_network(adj)

n = length(adj);
spaths=inf(n,n);
adjk = adj;

% calculate Number of shortest paths
for k=1:n-1

  for i=1:n
    for j=1:n

      if adjk(i,j)>0; spaths(i,j)=min([spaths(i,j) adjk(i,j)]);
end

    end
  end

  adjk=adjk*adj;

end


betw = zeros(1,n);
for i=1:n
    [dist,P]=dijkstra(adj,i,[]);
    for j=1:n

        if dist(j)<=1; continue; end   % either i=j or i,j are 1
edge apart
        betw(P{j}(2:dist(j))) = betw(P{j}(2:dist(j))) +
1/spaths(i,j);

    end
end

betw=betw/nchoosek(n,2);   % further normalize by the Number of
all node pairs

figure;
 stem(betw)
 xlabel('node');
 ylabel('betweenness centrality of node');
 title('Value of betweenness centrality of a Network node');
```

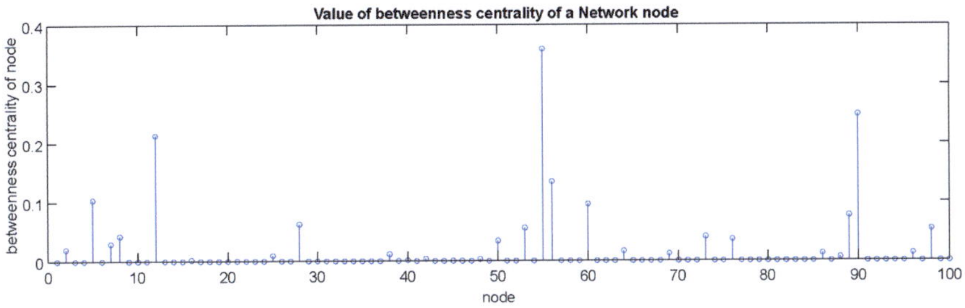

Fig. 4.14 Betweenness centrality of network consisting of 100 nodes

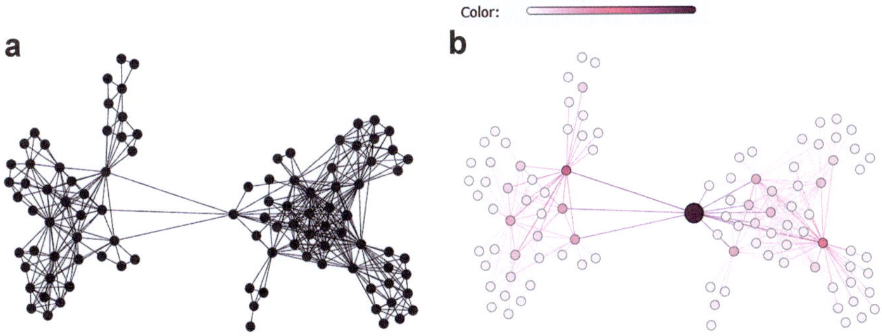

Fig. 4.15 (**a**) The network and (**b**) betweenness centrality of network consist of 100 nodes. The color map indicates the intensity of the measure. The darker the node, the higher the value, and the node with the largest value of the clustering coefficient has a larger size

For the calculation in the Matlab, we type the following command in the corresponding window. The corresponding results for each node of the network appear in Fig. 4.14.

```
>> betweenness_centrality_network(Net_A1)

% Net_A1 is the name of adjacency matrix
```

In Fig. 4.15, the network is shown on the left, while on the right is the network layout with a color scale based on the betweenness centrality of each node. A bright purple color in the case of node $N = 55$ is employed to represent the highest value and thus the larger circle size. It is also visually perceptible according to the importance of the variable that this node will have the highest value as we can see that it is on a critical path. Such nodes are usually called hubs in analogy with the airport hubs (i.e., airports presenting connections with many other airports, not necessarily connected directly between them).

4.2.5 Closeness Centrality

Closeness centrality represents the flow of information from one node to others and measures how short the shortest paths are from node i to all nodes. The closeness centrality can be calculated using the following equation:

$$\mathrm{CC}(i)=\frac{N-1}{\sum_{j} d(i,j)} \tag{4.8}$$

where $i \neq j$, d_{ij} is the length of the shortest path between nodes i and j in the network and N is the number of nodes.

The closeness coefficient can be calculated using script 4.7.

```
% Script 4.7
% Example of network closeness coefficient

function [ Cc ] = closeness_centrality_network(adj)

%   INPUTS      adjacency  matrix  adj

%   OUTPUTS     a vector with values of closeness centrality of
each node


ADJ=adj;
ADJ(ADJ>0)=1;
n=size(ADJ);
ADJ(ADJ==0)=NaN;
D=distance_wei(ADJ);
D(isinf(D))=NaN;
if n(1)~=n(2);
    'The input matrix does not refer to a graph. Calculations
cannot be done'
else
    n=n(1);
    Cc=(n./nansum(D))';
end
Cc(isinf(Cc))=NaN;

figure;
 stem(Cc)
 xlabel('node');
 ylabel('closeness coefficient of node');
 title('Value of Closeness Coefficient of a Network node');
```

For the calculation, we type in the Matlab command window the following command:

```
>> closeness_centrality_network(Net_A1)

%  Net_A1 is the name of adjacency matrix
```

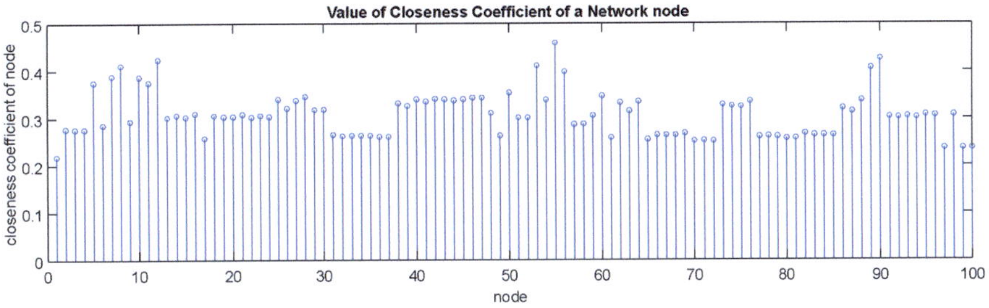

Fig. 4.16 Closeness centrality of network consists of 100 nodes

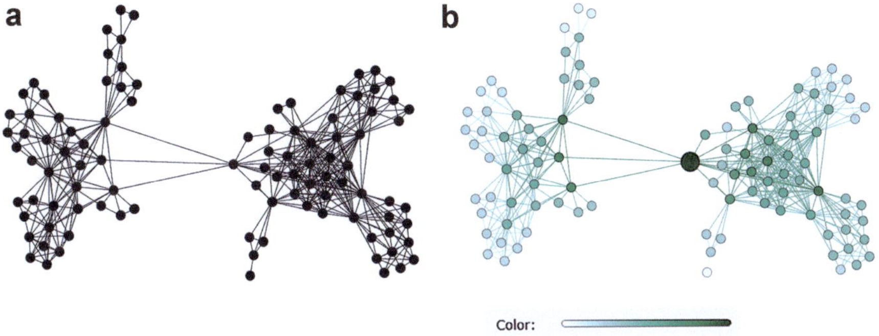

Fig. 4.17 (**a**) The network and (**b**) closeness centrality of nodes of a network consisting of 100 nodes. Nodes with a bright green color and a larger circle size than the rest are the nodes that present a higher closeness centrality value than the others

The result of the calculation is illustrated in Fig. 4.16 where we can see the closeness centrality for each node of the network.

Figure 4.17 shows in the left the network and in the right their representation in color scale based on the value of closeness centrality of each node. Nodes with a bright green color and a larger size than the rest are the nodes with a greater closeness centrality value than the others.

4.2.6 *Eigenvector Centrality*

Eigenvector centrality is a measure of a node's influence within a network while considering the importance of its neighbors. This measure takes into account not only how many connections a node has (i.e., its degree), but also the centrality of the vertices that it is connected to. The eigenvector centrality is based on the eigenvalue, meaning that the value of an entity is based on the value of the entities connected to it: the higher the latter is, the higher the former becomes.

The following script (script 4.8) permits to calculate the eigenvector centrality in Matlab.

```
% Script 4.8
% Example of network eigenvector centrality

function    v = eigenvector_centrality_network(adj)


%   Inputs     adj          adjacency matrix.
%
%   Outputs     v          eigenvector centrality


n = length(adj) ;
if n < 1000
     [V,~] = eig(adj) ;
     ec = abs(V(:,n)) ;
else
     [V, ~] = eigs(sparse(adj)) ;
     ec = abs(V(:,1)) ;
end
v = reshape(ec, length(ec), 1);
figure;
 stem(v)
 xlabel('node');
 ylabel('eigenvector centrality of node');
 title('Value of eigenvector centrality of a Network node');
```

In the Matlab command window, we type:

```
>> eigenvector_centrality_network(Net_A1)

%  Net_A1 is the name of adjacency matrix
```

Figure 4.18 shows the values of the indicator for each node, while Fig. 4.19b shows the network and its colored representation based on the value of eigenvalue centrality, where we can see in bright green the nodes with a higher value.

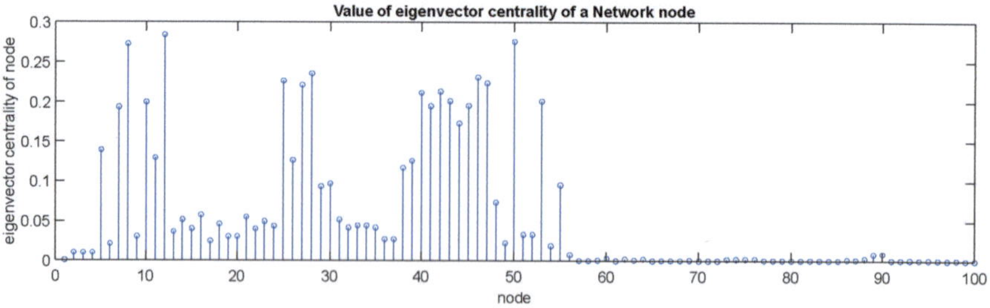

Fig. 4.18 Eigenvector centrality of the nodes of network consisting of 100 nodes

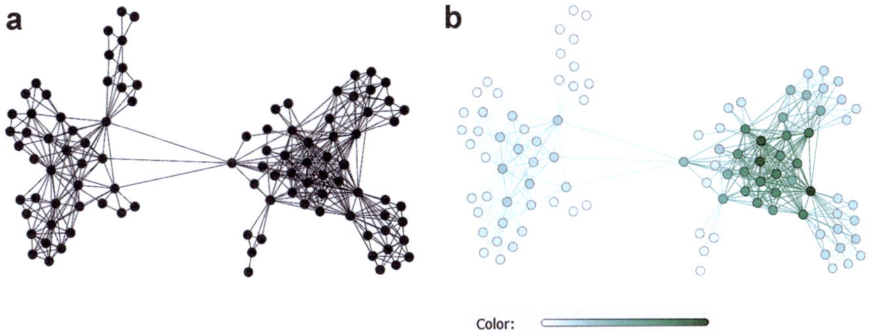

Fig. 4.19 (**a**) The network and (**b**) Eigenvector centrality of network consist of 100 nodes. The color variation indicates the value of the measure. The darker the color, the higher the value

4.2.7 *Modularity*

Modularity measures the effectiveness of a network's division into communities. Communities with high modularity values have dense edge connections between the vertices within a community, but sparse connections between nodes in different communities.

We can use the script 4.9 to calculate modularity of network in Matlab.

```
% Script 4.9
% Example of network modularity

function [Mi Mv]=modularity_network(adj)

%   Inputs     adj         adjacency matrix
%
%   Outputs    Mi           modularity vector
```

In Matlab command window, we can type the following commands:

```
>> modularity_network(Net_A1)

% Net_A1 is the name of adjacency matrix
```

In Fig. 4.20, the results are presented, and we see the modularity values for each node. Figure 4.21 shows these groups of nodes in different colors. We can distinguish six different groups of nodes (modules), which are depicted in different colors. Within these groups, there are strong connections between nodes and sparse connections between groups (modules).

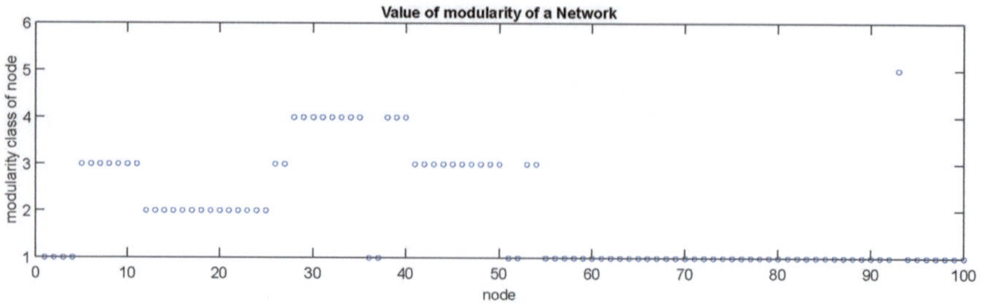

Fig. 4.20 Modularity class of each node of the network

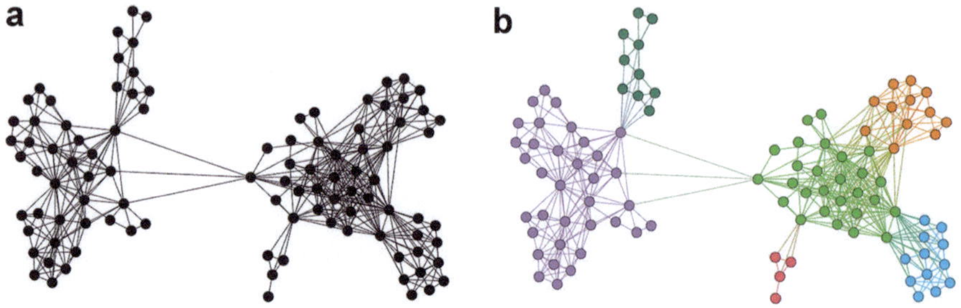

Fig. 4.21 (**a**) The network and (**b**) modularity of network consisting of 100 nodes. The different colors represent nodes in different communities

4.3 Types of Networks

A network is essentially a system that can be represented as a graph, consisting of elements known as nodes or vertices and a set of connecting links (edges) that represent the interactions between them [1, 7, 8, 9, 13].

4.3.1 Small-World Network

A small-world network is a type of graph where most nodes are not directly connected to each other, but most nodes can be reached from every other node within few steps. A small-world network is a network where the typical distance L between two randomly chosen nodes grows proportionally to the logarithm of the number of nodes N in the network, $L \propto \log N$. Watts and Strogatz introduced this type of network as follows [13]:

Consider a set of n vertices $\{v_1, v_2, \ldots, v_n\}$ and an (even) number k. In order to ensure that the graph will have relatively few edges (i.e., it is sparse), choose n and k such that $n \geq k \geq \ln(n) \geq 1$.

- Order the n vertices into a ring and connect each vertex to its first $k/2$ left-hand (clockwise) neighbors, and to its $k/2$ right-hand (counterclockwise) neighbors, leading to graph G

- With probability p, replace each edge $\{u, v\}$ with an edge $\{u, w\}$ where w is a randomly chosen vertex from $V(G)$ other than u, and such that $\{u, w\}$ is not already contained in edge set of (the modified) G

Watts–Strogatz model presents high clustering coefficients while maintaining short average path lengths. The resulting graph is known a Watts–Strogatz random graph or WS graph.

To make a Watts–Strogatz (WS) graph, we start with a ring lattice. First, a ring lattice with N nodes of mean degree 2K will be created. Each node is connected to its K nearest neighbors on either side. Next, for each edge in the graph, rewire the target node with probability p. We can employ script 4.10 to create a WS network in Matlab.

```
% Script 4.10
% Example of Watts-Strogatz (WS) network

function WS = WattsStrogatz(N,K,p)

% N number ofnodes, K degree and p the probability.
%
% When p = 0 is a ring lattice, and p = 1 is a random graph.

s = repelem((1:N)',1,K);
t = s + repmat(1:K,N,1);
t = mod(t-1,N)+1;

for source=1:N
    SE = rand(K, 1) < p;

    NT = rand(N, 1);
    NT (source) = 0;
    NT (s(t==source)) = 0;
    NT (t(source, ~ SE)) = 0;

    [~, ind] = sort(NT, 'descend');
    t(source, SE) = ind(1:nnz(SE));
end

WS = graph(s,t);
end
```

In order to generate a network, consist of 20 nodes, a network degree of 5 and probability 0, we write in Matlab the following commands.

```
% Example of Watts-Strogatz (WS) network

>> WS = WattsStrogatz(20,5,0);

>> plot(WS,'NodeColor','k','Layout','circle');

>> title('Watts-Strogatz Graph with N = 20 nodes, degree K = 5, and
p = 0')
```

The results are presented in Fig. 4.22. The graph is a perfect ring lattice. When $p = 0$, no edges are rewired and the model returns a ring lattice. Conversely, when $p = 1$, all of the edges are rewired, converting the ring lattice is transformed into a random graph.

Then we increase the probability to 0.1 and we type in Matlab the following commands. The results are presented in Fig. 4.23.

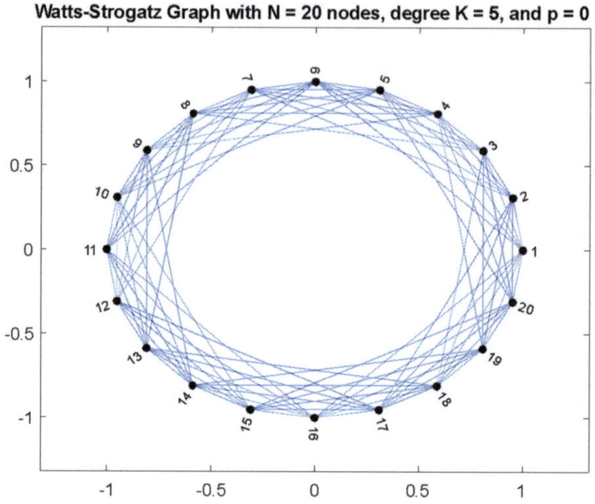

Fig. 4.22 A perfect lattice network with 20 nodes

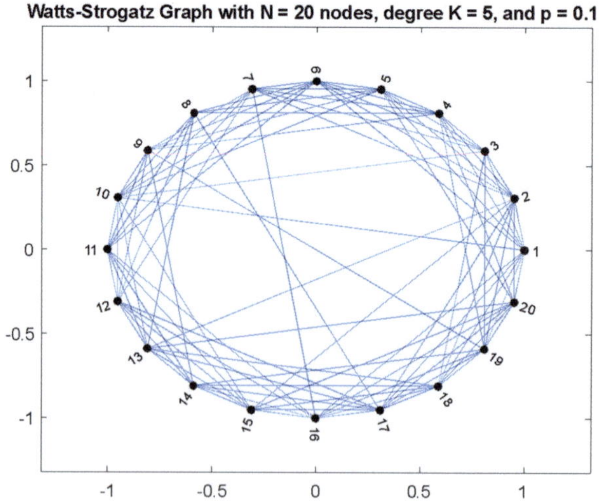

Fig. 4.23 Network with 20 nodes and $p = 0.1$

```
% Example of Watts-Strogatz (WS) network

>> WS = WattsStrogatz(20,5,0.1);

>> plot(WS,'NodeColor','k','Layout','circle');

>> title('Watts-Strogatz Graph with N = 20 nodes, degree K = 5, and
p = 0.1')
```

Then we increase the probability to 1 by typing the command, and the result is presented in Fig. 4.24. As we can see by comparison to Fig. 4.23, as probability increases, the number of random connections increases too.

```
>> WS = WattsStrogatz(20,5,1);

>> plot(WS,'NodeColor','k','Layout','circle');

>> title('Watts-Strogatz Graph with N = 20 nodes, degree K = 5, and
p = 1')
```

The degree distribution of the nodes in the different Watts–Strogatz graphs varies. When beta is 0, all nodes preset the same degree, 2K. However, as p increases, the degree distribution changes. This can be seen in Fig. 4.25.

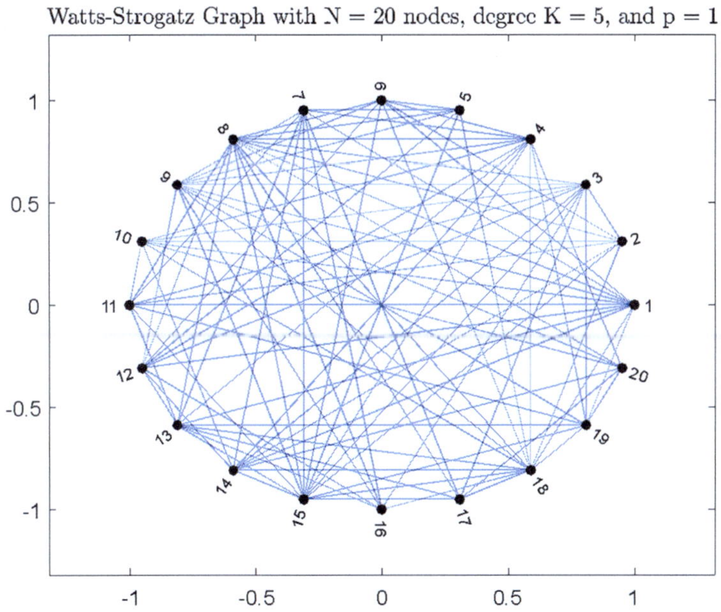

Fig. 4.24 Random network with 20 nodes

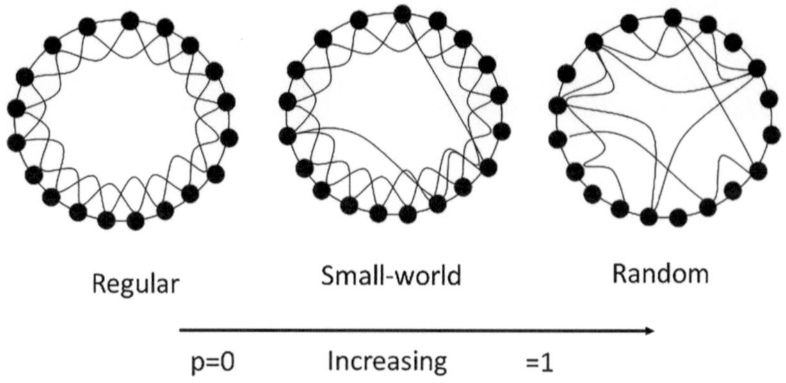

Fig. 4.25 The Watts–Strogatz model of the small world. For $p = 0$, the network is regular and as it increases, it is characterized as small world until for $p = 1$ where it is a random graph

4.3.2 Scale-Free Network

A scale-free network is one in which the distribution of links to nodes follows a power law. The power law means that the vast majority of nodes have very few connections, while a few important nodes (named Hubs) have a huge number of connections.

Albert-László Barabási mapped the network of a portion of the World Wide Web (WWW). The analysis of that network had led to some interesting findings:

1. A number of nodes (hubs) have more connections than others.
2. The WWW network has a power law distribution of the number of links connected to web pages.

From the above, we can conclude that scale-free networks have the following key features:

1. Several nodes with high degrees are known as hubs; they appear as a result of preferential attachment.
2. The degree distribution follows a power law.
3. Hubs usually have links from all around the network, serving as links between different parts of the network, therefore showing a small-world property.

Figure 4.26 represents an example of a network with scale-free behavior and the corresponding degree distribution in Fig. 4.27, which was generated by running the commands of script 4.11.

```
%Script 4.11

% Suppose we have a network name SFN275

Inp1=input('name of adjacency_');

Degree_Distribution_net(Inp1)
```

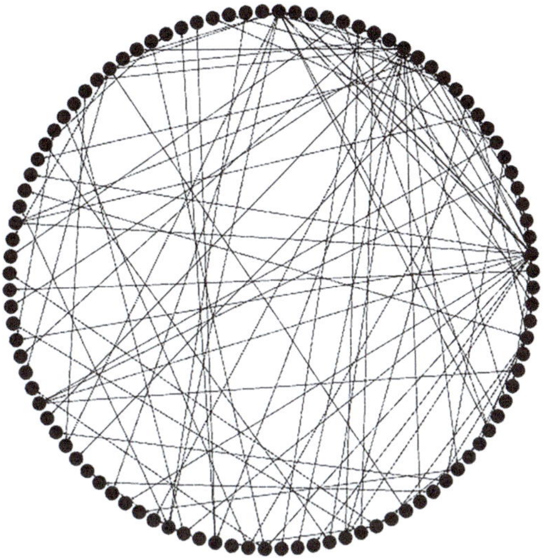

Fig. 4.26 Scale-free network

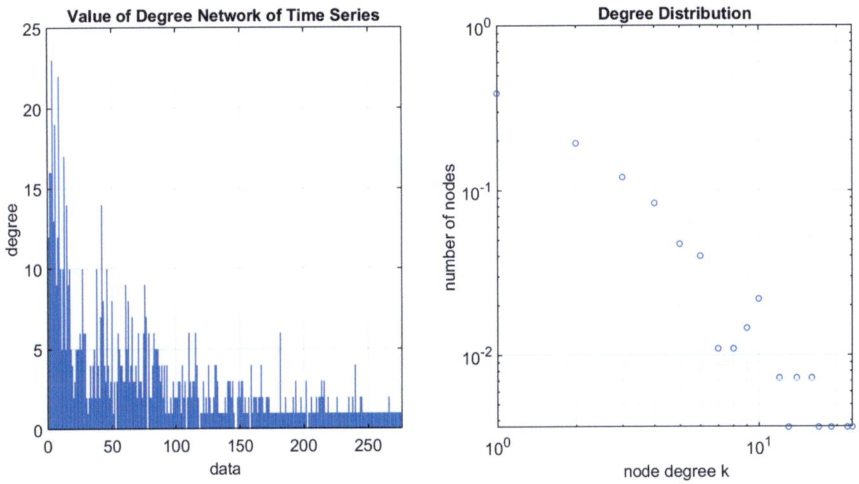

Fig. 4.27 Degree and degree distribution of scale-free network of Fig. 4.26

4.3.3 Random Network

A random network consists of N nodes where each node pair is connected with probability p [8]. To construct a random network, we follow these steps:

1. Start with N isolated nodes.
2. Select a node pair and generate a random number between 0 and 1.

If the number is larger than the probability p, then connect the selected node pair with a link otherwise leave them disconnected.

3. Repeat step 2 for each of the $N(N-1)/2$ node pairs.

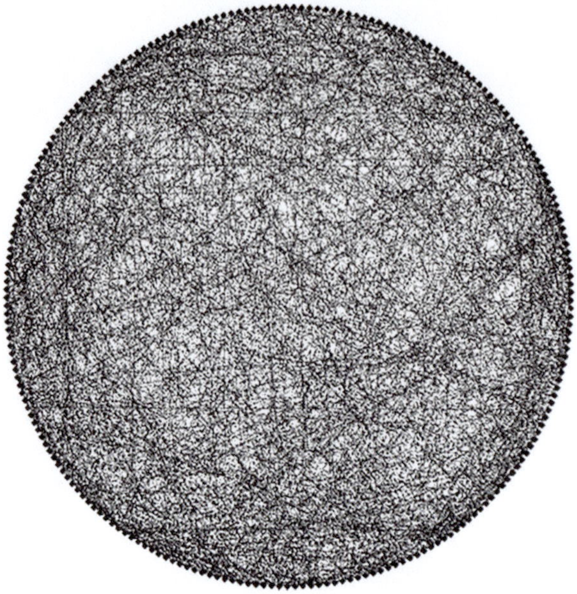

Fig. 4.28 Random network consists of 225 nodes

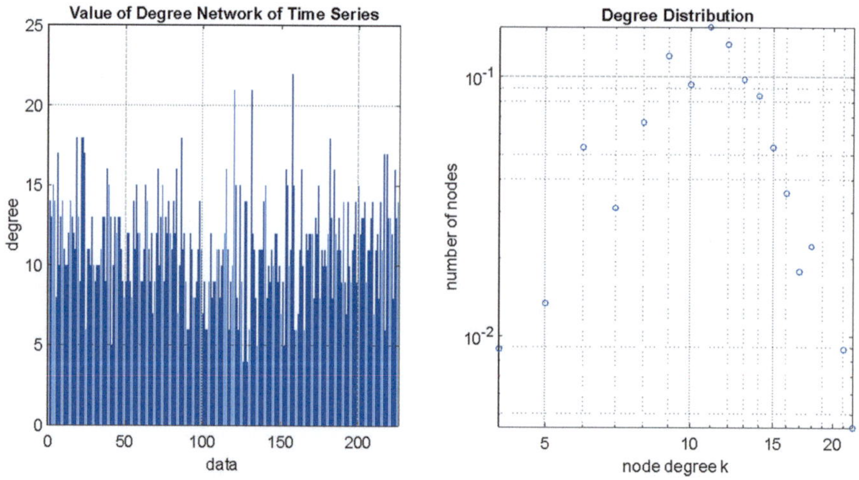

Fig. 4.29 Degree and Degree Distribution of the Random network of Fig. 4.28

The network obtained after this procedure is called a random graph or a random network. A representative example is presented in Fig. 4.28 along with the degree value for each node and the degree distribution in Fig. 4.29.

Random networks' degree distribution follows either a Poisson or Binomial distribution.

4.4 From Time Series to Complex Network/Methods of Construction

The concept of analyzing the dynamic characteristics of a time series by transforming it into a complex network system has been introduced [1, 6, 9, 10, 12]. Several studies have shown that distinct features of a time series can be mapped onto networks with different topological characteristics such as correlation, phase space reconstruction, visibility, and recurrence.

4.4.1 Phase Space Network: Recurrence Network

Recurrence networks are a technique used to analyze complex dynamical systems based on the recurrence properties of states in the system's phase space. They are particularly useful for understanding the underlying dynamics of nonlinear and chaotic systems. Recurrence networks offer a graph-based representation of time series data, providing insights into the system's recurrent patterns and structures. Recurrence networks offer a unique perspective on time series data by emphasizing the spatial relationships between states in a reconstructed phase space. They are a valuable tool for researchers seeking to understand the complex behaviors of dynamic systems.

First, recurrence networks start with a time series dataset, which consists of a sequence of data points recorded over time. This data can originate from various domains, such as physics, biology, finance, and engineering just to mention few ones. Then, to create a recurrence network, we first transform the one-dimensional time series data into a higher-dimensional representation known as a phase space. This is often done using time delay embedding, a technique that constructs a multi-dimensional space by stacking lagged copies of the time series. The choice of embedding dimension and time delay is important and should be determined based on the characteristics of the system under study.

The core of a recurrence network is the construction of a recurrence plot. A recurrence plot is a binary matrix where each element indicates whether two points in the phase space are close to each other. Typically, a threshold or distance measure is used to determine when two points are considered close or "recurrent." If the distance between two points falls below the threshold, the corresponding entry in the recurrence plot is set to 1; otherwise, it's set to 0. Recurrence plots reveal regions in phase space where the system revisits similar states over time. Finally, the recurrence plot can be further transformed into a network or graph. In this network, nodes represent the states in the phase space, and edges connect pairs of states that are recurrent. The edges can be weighted, representing the strength of recurrence between states. Recurrence networks can be applied to a wide range of dynamic systems, including ecological systems, physiological data (e.g., EEG signals), climate data, financial time series, and more. They have found applications in fields like physics, biology, engineering, and neuroscience.

Advantages of recurrence networks include their ability to capture complex, nonlinear dynamics and their robustness to noise in the data. They can reveal important features of a system's behavior that may not be apparent through traditional time series analysis techniques.

Overall, recurrence networks provide a powerful tool for understanding the underlying structure and behavior of complex systems, making them valuable in various scientific and engineering disciplines.

4.4.2 Correlation Network

A correlation network is a type of network that is constructed based on the statistical relationships between variables or data points, specifically through measures of correlation. These networks are used to visualize and analyze the associations or dependencies among variables in a dataset. Correlation networks are particularly useful for understanding patterns and connections in multivariate data.

In this approach, we calculate the pairwise correlations between time series data points (e.g., Pearson correlation coefficient or cross-correlation) and use these correlations to construct a network. Nodes in the network represent time series variables, and edges between nodes represent significant correlations. Such networks can help identify which variables are strongly related over time.

To create a correlation network, we start with a dataset that contains multiple variables (features) and data points. These variables can represent anything from financial indicators, biological measurements, to social interactions, and more. Then, the next step is to calculate pairwise correlations between all pairs of variables in the dataset. The most commonly used correlation measure is the Pearson correlation coefficient, but other measures like Spearman's rank correlation or Kendall's tau can also be used, depending on the nature of the data.

Once we have computed the correlations, we typically apply a threshold to decide which correlations to include in the network. Correlations above a certain threshold (e.g., absolute correlation value >0.5) are considered significant and are used to establish connections in the network. The choice of threshold can impact the network's density and structure.

Based on the significant correlations, we construct a network where each variable is represented as a node (or vertex), and edges (or links) between nodes represent correlations above the threshold. The edges can be weighted to reflect the strength of the correlation.

The resulting correlation network can be visualized using various techniques, such as node-link diagrams or adjacency matrices. Visualization tools like network graphs can help you explore and interpret the relationships between variables.

Correlation networks have applications in various fields like finance, biology, climate science, and other.

4.4.3 Visibility Network

We remind that the structure of complex network can be represented as a graph $G = (N,E)$, which consists of a set of $N = (n_1,n_2,\ldots,n_N)$ vertices or nodes connected by a set of $E = (e_1,e_2,\ldots,e_E)$ links or edges. A network can be represented by its adjacency matrix $A = [a_{ij}]$. The adjacency matrix contains the information about the connectivity structure of the graph, and for a graph with N nodes is an $N \times N$ matrix. The elements a_{ij} are equal to 1 whenever there is an edge connecting the vertices i and j, and equal to 0 otherwise. When the graph is undirected, the adjacency matrix is symmetric, i.e., the elements $a_{ij} = a_{ji}$ for any i and j.

In the visibility method, each value of time series is converted to a node and each node is connected with all the other nodes that exists visibility between them. There are two main categories and several variations of them, converting time series to network, horizontal graph, and natural graph. The natural graph is described below.

Mathematically, the visibility criterion can be defined as follows: two time series points $x(t_i)$ and $x(t_j)$ in the time series have visibility and consequently become two connected nodes in the graph, if any other data $(t_k, x(t_k))$ placed between them $(t_i < t_k < t_j)$ fulfills the following constrain:

$$x(t_k) < x(t_i) + \left(x(t_j) - x(t_i)\right)\frac{t_k - t_i}{t_j - t_i} \tag{4.9}$$

Hence, i and j are connected if a straight line can be drawn in the time series joining the two points i and j, such that, at all intermediate points $(t_i < t_k < t_j)$, $x(t_k)$ fall below this line. In a network mapped using the visibility algorithm, each node is visible at least by its nearest neighbors (left and right). An illustration of a time series transformed into a visibility graph is shown in Fig. 4.30.

To understand the methodology of converting time series to network, below we present an example of converting a time series of ten values to a network and a time series of 10 points to a network (Fig. 4.31).

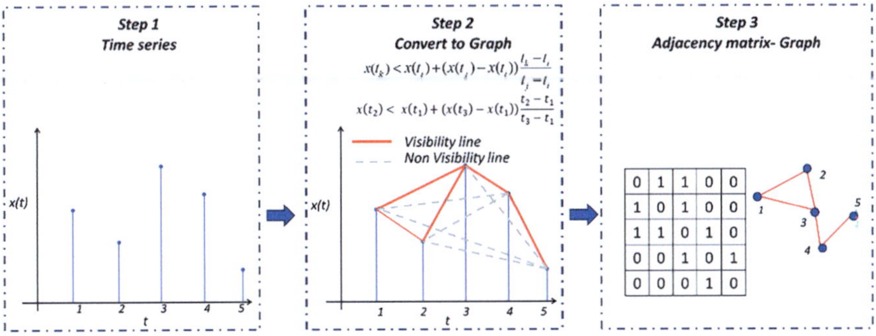

Fig. 4.30 Schematic representation of transformation of time series to a network via the visibility method

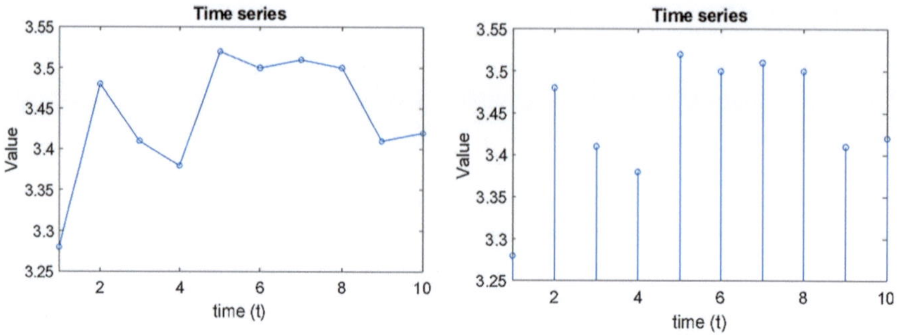

Fig. 4.31 Example of time series with ten values (two different plot type)

Then the algorithm for converting a time series into a graph is based on the visibility algorithm as presented in script 4.12.

```
%Script 4.12

% Example of visibility transformation

function [Net]=visibilitynet(xn)

% xn the name of time series

N=length(xn);
txV=[1:1:N]';
xV=xn;
for i=1:N
    if (i<N)
        Net(i,i+1)=1;
        Net(i+1,i)=1;
    end
end
for i=1:N
    for j=(i+2):N
        Dyt(i,j)=(xV(j)-xV(i))/(txV(j)-txV(i));
        Net(i,j)=1;
        Net(j,i)=1;
        for k=(i+1):(j-1)
            temp(k) = xV(i) + Dyt(i,j)*(txV(k)-txV(i));
            if temp(k) <= xV(k)
                Net(i,j) = 0;
                Net(j,i) = 0;
                break
            end
        end
    end
end
```

Suppose we have NetA1_10 time series, and we want to transform it into a network named NETA1_10. We have to write the following in Matlab (Fig. 4.32).

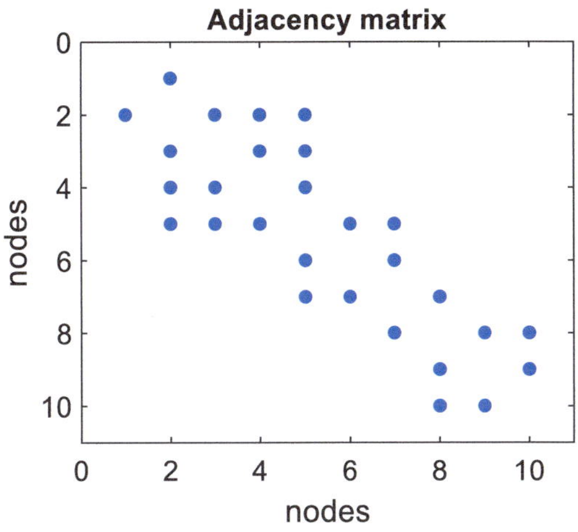

Fig. 4.32 Adjacency matrix of time series with ten values (two different plot type)

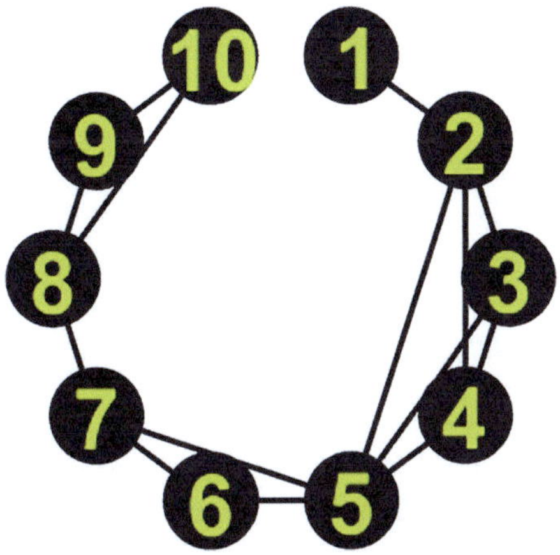

Fig. 4.33 Network of time series with ten values

```
>> NETA1_10=visibilitynet(NetA1_10)

% NETA1_10 is the name of the network

% NetA1_10 is the name of the time series
```

As we have already seen for the adjacency matrix wherever there are dots, it means that there is a connection between the corresponding nodes. That is, there are a total of ten nodes as many as the points of the time series. Connections are derived from the criterion of visibility graph. In the adjacency matrix, where there is a dot means connection between the nodes. That is, we can see that node 1 is connected only to node 2, while node 2 is connected to node 1, 3, 4, 5. In Fig. 4.33, the graph formed by the adjacency matrix is shown.

4.5 Extended Example of Transforming Time Series to Network and Analyzing Them Using Network Properties

4.5.1 Examples of Field Measurement Data (Environmental Time Series)

Below we will present some examples of converting time series into networks. The time series have been selected on the base of their different dynamical behavior. Specifically, these are field meteorological time series, which have been obtained through the Poseidon system of buoys maintained by the Hellenic Center for Marine Research (HCMR) www.poseidon.hcmr.gr. The sampling interval was 3 h for all variables, and every measurement is an average over 10 min, resulting in eight values per day. Thus, the length of the time series is 3.730 values (in units of $\Delta t = 3$ h).

At first, time series of water temperature (Water Temp.) is displayed in Fig. 4.34. In order to plot the time series, we write the following command.

```
First example of field time series

% Suppose the name of time series is Water Temp

% Plot time series

>> plot(Water Temp)
```

In order to convert this time series into a network using the visibility method, we write the following command in Matlab. We remind that in this way, the number of nodes of the network is equal to the number of the time series recordings (3730 in the present case).

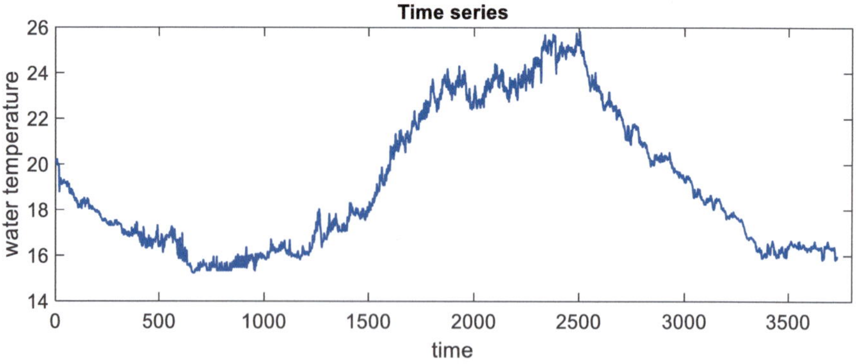

Fig. 4.34 Water temperature time series

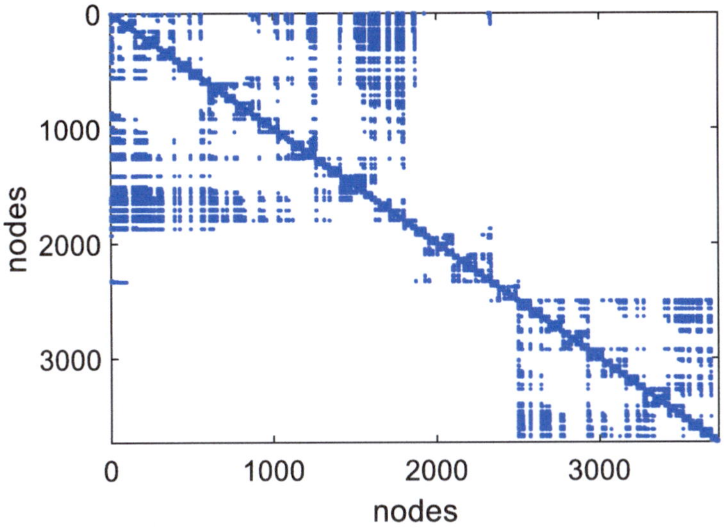

Fig. 4.35 Adjacency matrix of water temperature time series of Fig. 4.34

```
First example of field time series

% Suppose the name of time series is Water_temp
% Net_Water_temp the given name of network

>> Net_Water_temp=visibilitynet(Water_temp)
```

In this way, we have transformed the time series into a network. To see the adjacency matrix we write in Matlab (Fig. 4.35):

```
First example of field time series

% Suppose the name of time series is Water_temp
% Net_Water_temp is the name of network

>> spy(Net_Water_temp)
```

In order to be able to visualize the network in the design software Gephi [2], we write in Matlab the following command.

```
First example of field time series

% Net_Water_temp is the name of network
% Transform network to net format
% NET_WATER_TEMP the name of net format file

>> writetoPAJ(Net_Water_temp,'NET_WATER_TEMP',0)
```

Using the above routine, the file can now be processed by the Gephi program, the visualization of which based on modularity measure is shown in Fig. 4.36.

Looking at the network, we can see that few groups (modules) of nodes are created where nodes have strong connections within the same group and sparse connections of nodes between different groups (modules). This observation is linked to the dynamic evolution of time series values.

In the following, we present another example of a time series transformation into a network. After executing the following command in Matlab, the time series is shown in Fig. 4.37.

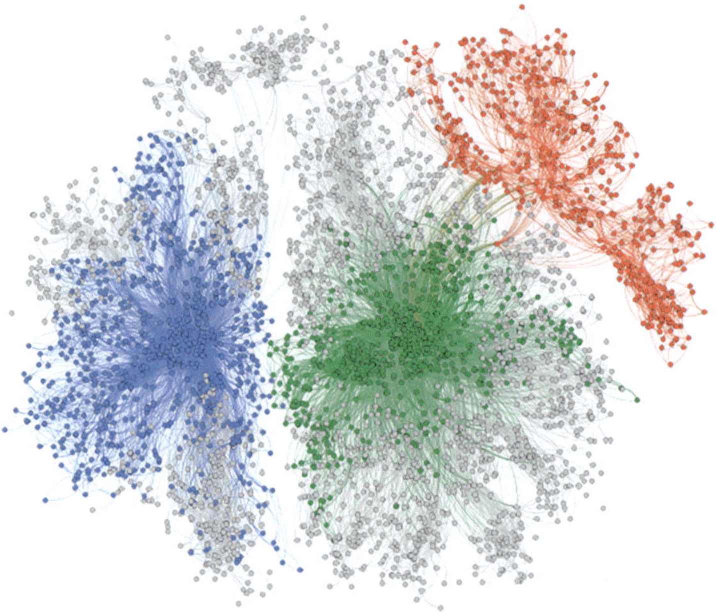

Fig. 4.36 Network of water temperature time series

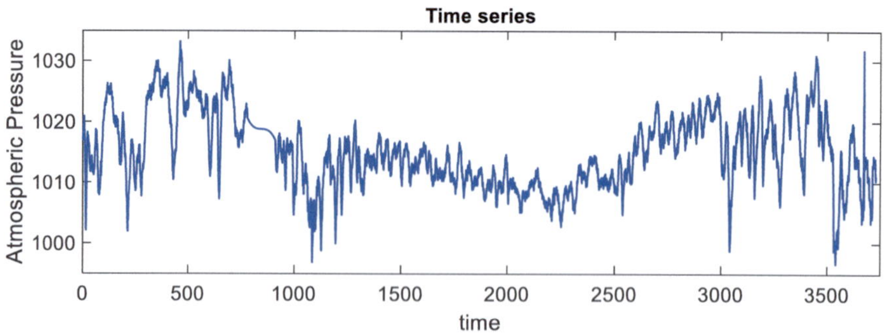

Fig. 4.37 Atmospheric pressure time series

```
Second example of field time series

% Suppose the name of time series is Air_Pressure

% Plot time series

>> plot(Air_Pressure)
```

We observe that the values of this time series have different dynamic behavior, where the periodicity of the data is no longer clear, and these values fluctuate more.

We convert this time series into a network by typing the following command in Matlab.

```
Second example of field time series

% Suppose the name of time series is Air_Pressure

% Net_Air_Pressure the given name of network


>> Net_Air_Pressure=visibilitynet(Air_Pressure)
```

We can see the adjacency table by writing in Matlab (Fig. 4.38).

```
Second example of field time series

% Suppose the name of time series is Air_Pressure

% Net_Air_Pressure is the name of network


>> spy(Net_Air_Pressure)
```

The next figure illustrates the network using the Gephi program, in the property of modularity.

In the network representation in Fig. 4.39, we can distinguish several groups of nodes where there are strong connections, and more sparse connections between groups.

The third example of converting a time series to a network involves a time series that fluctuates more than the two previous one. After executing the following command in Matlab, in Fig. 4.40 the time series is shown.

```
Third example of field time series

% Suppose the name of time series is Wind_speed

% Plot time series

>> plot(Wind_speed)
```

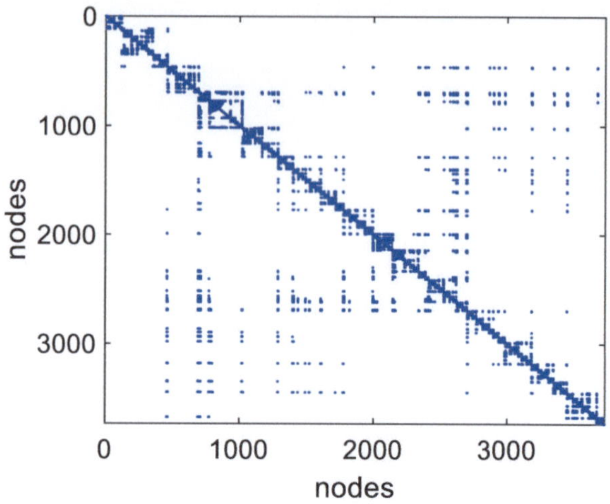

Fig. 4.38 Adjacency matrix of atmospheric pressure time series of Fig. 4.37

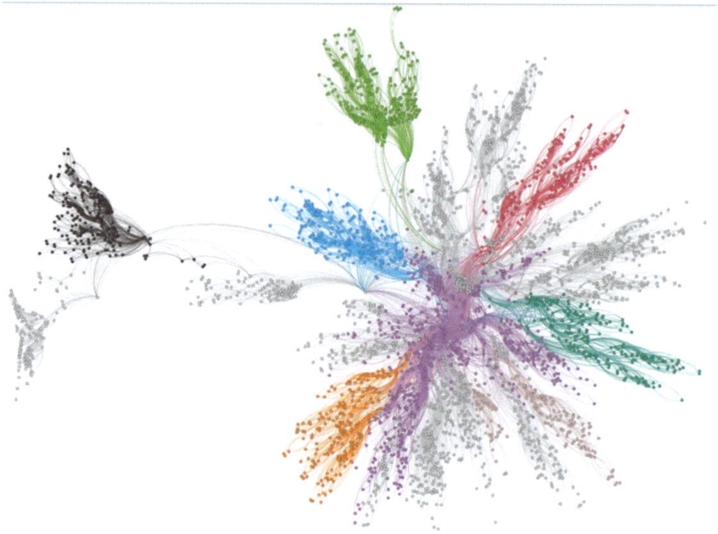

Fig. 4.39 Network of atmospheric pressure time series

To convert this time series into a network using the method of visibility, we write the following command in Matlab.

```
Third example of field time series

% Suppose the name of time series is Wind_speed

% Net_Wind_speed the given name of network


>> Net_Wind_speed=visibilitynet(Wind_speed)
```

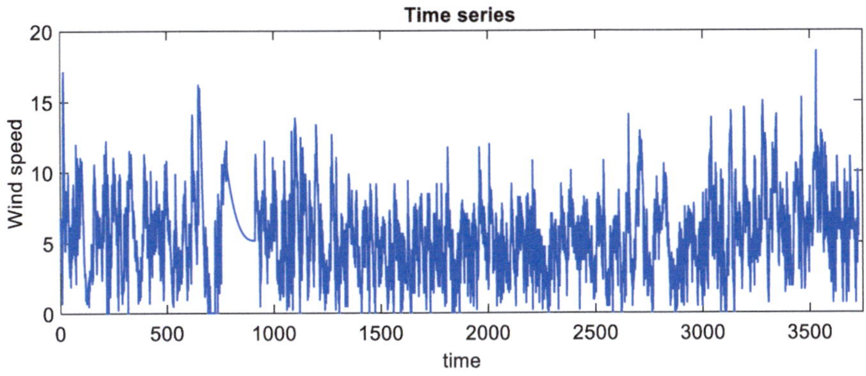

Fig. 4.40 Wind speed time series

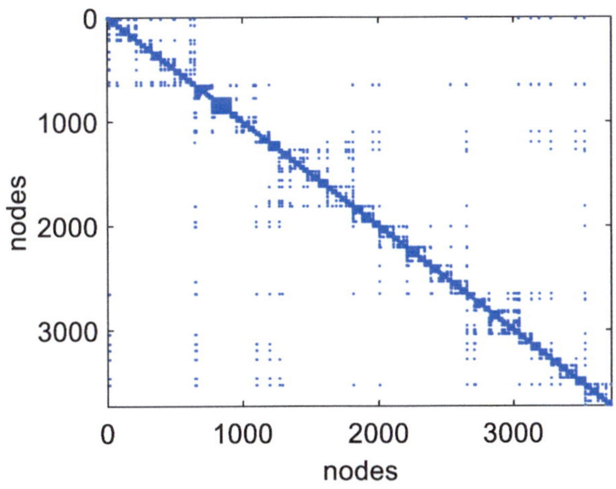

Fig. 4.41 Adjacency matrix of wind speed time series of Fig. 4.40

In this way, we have transformed the time series into a network. To obtain and visualize the adjacency matrix, we write in Matlab (Fig. 4.41):

```
Third example of field time series

% Suppose the name of time series is Wind_speed

% Net_Wind_speed is the name of network


>> spy(Net_Wind_speed)
```

To visualize the network in the design software Gephi, we write the following command in Matlab.

```
Third example of field time series
% Net_Wind_speed is the name of network
% Transform network to net format
% NET_WIND_SPEED the name of net format file
>> writetoPAJ(Net_Wind_speed,'NET_WIND_SPEED',0)
```

In Fig. 4.42, the corresponding network is shown, using the modularity measure for illustration.

It is appropriate to present the three cases in the figure below and then comment on the network analysis method (Fig. 4.43).

In the left part, we plot the variable as recorded, in the middle, we present the network's adjacency matrix, and on the right, we represent the network graph, where we mention the modularity classes.

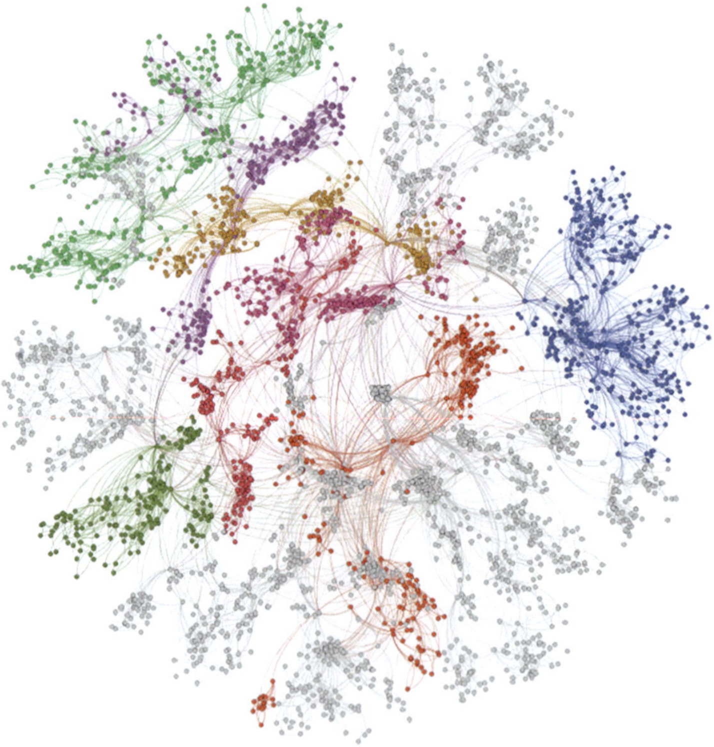

Fig. 4.42 Network of wind speed time series

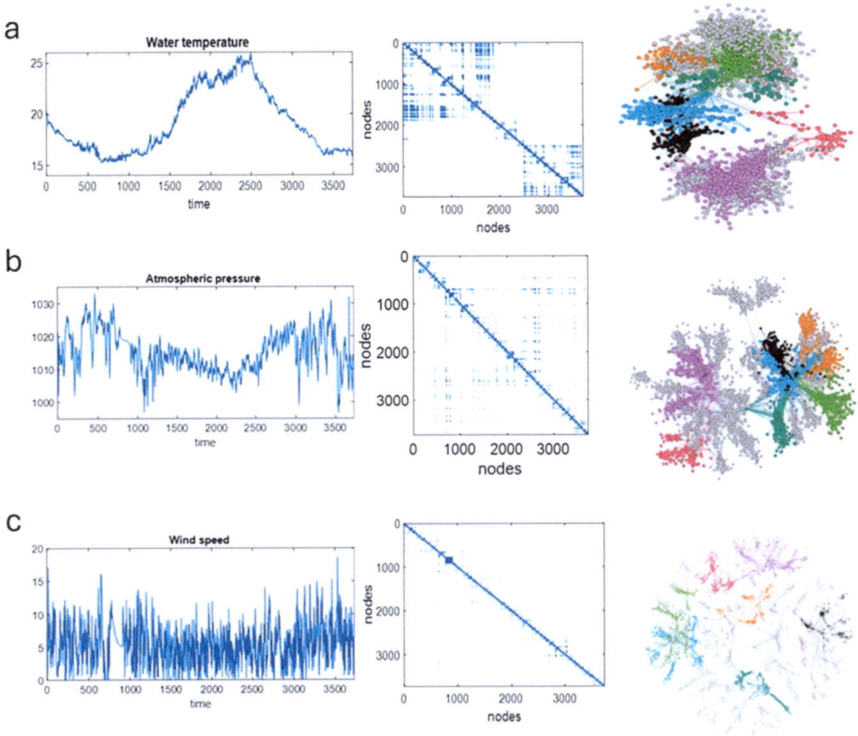

Fig. 4.43 Time series, adjacency matrix, and network graph generated, respectively, for (**a**) air temperature, (**b**) water temperature, and (**c**) wind speed time series using the visibility algorithm

A first overview suggests that differences in the dynamical behavior of the observations are mapped onto corresponding network topologies. These differences are associated with the fluctuations of the time series. Specifically, in the water temperature network, there are relatively few communities (modularity classes) each containing a large number of connected nodes. In contrast, the air temperature network exhibits a greater number of communities compared to those of water temperature network with a corresponding reduction of the connected nodes in each hub. Finally, the wind speed network presents a lot of communities with the least connected nodes in each hub. This network profile is consistent with the physical state, as it increases the fluctuation of the time series values. Hence, the network graphs provide an initial indication that the dynamic variability of the time series is effectively captured through network topology.

Below, we present in detail the calculation of the network's topological measures derived from the time series of water temperature.

To calculate the degree of the network and the distribution of the degree, we write in the command line of Matlab.

```
%Calculation of network degree of each node

% Net_Water_temp is the name of adjacency matrix that we have
calculated previous

% D_Wat_temp the degree  of network

>>D_Wat_temp=Degree_network(Net_Water_temp)
```

The result is reflected in Fig. 4.44 which shows the time series of the degree of each node of the network.

```
%Calculation of network degree and degree distribution

%  Net_Water_temp  is  the  name  of  adjacency  matrix  that  we  have
                                                   calculated previous
% D_Wat_temp the degree  of network

>>c

   Aver_DED = 18.0584
```

Below we present the degree and the degree distribution (Fig. 4.45).

The calculation of the diameter and the clustering coefficient are done by executing the following commands in Matlab.

```
%Calculation of network diameter

%  Net_Water_temp  is  the  name  of  adjacency  matrix  that  we  have
                                                   calculated previous

>> Diam_Water_temp=Diameter_Net(Net_Water_temp)

ans= 6
```

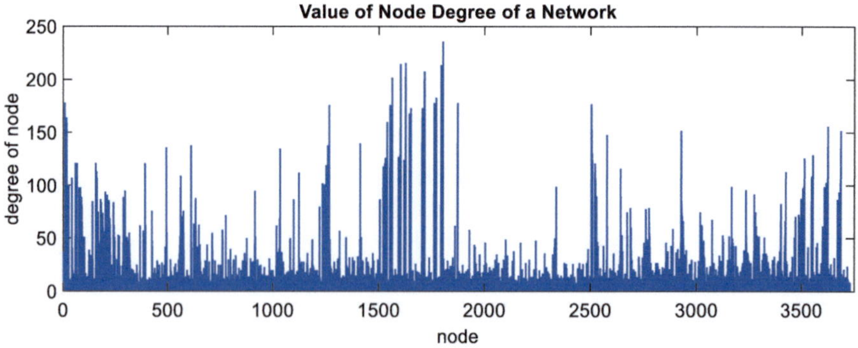

Fig. 4.44 Degree of water temperature network

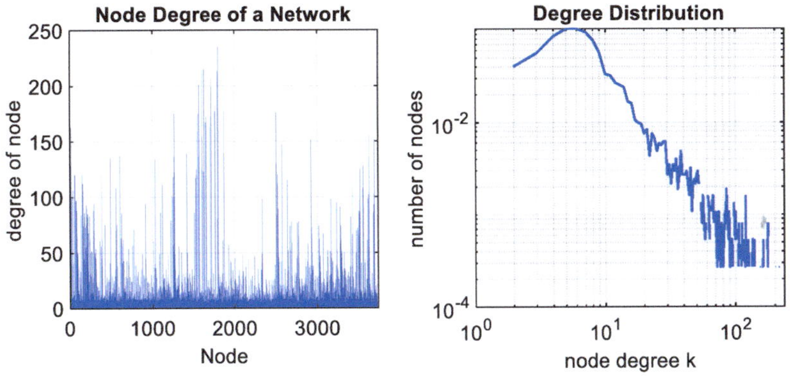

Fig. 4.45 Degree (left) and degree Distribution (right) of network of water temperature

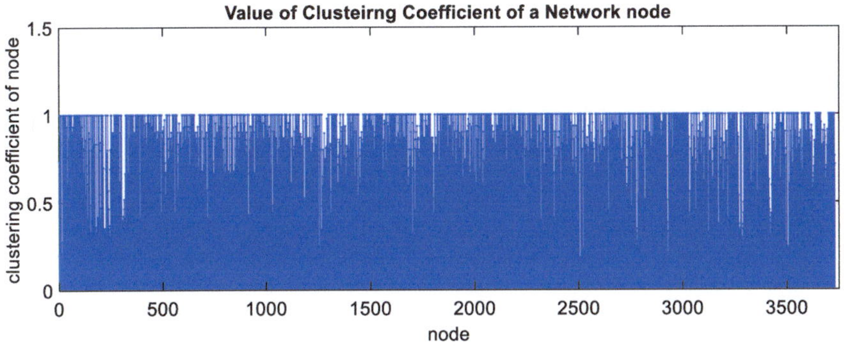

Fig. 4.46 Clustering coefficient of network of water temperature

```
%Calculation of network clustering coefficient

% Net_Water_temp is the name of adjacency matrix that we have
                                          calculated previous

>> [Cl, avCl_Water_temp]=clustering_coef_network(Net_Water_temp);

%   The clustering coefficient is the fraction of triangles around
a node

%   Input:     adj       adjacency matrix
%
%   Output:     Cl       clustering coefficient vector per node

%               avCl_Water_temp  mean of clustering coefficient

avCl_Water_temp  =0.6954
```

Figure 4.46 shows the clustering coefficient of water temperature network.

As we have mentioned, the betweenness centrality is calculated by typing the following command, and the result is represented in Fig. 4.47.

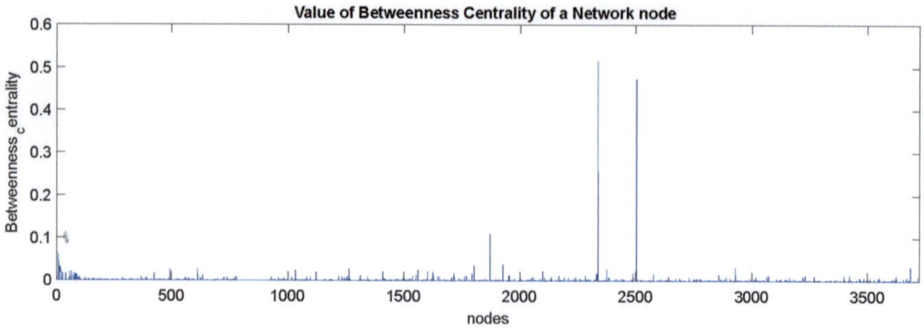

Fig. 4.47 Betweenness centrality of network of water temperature

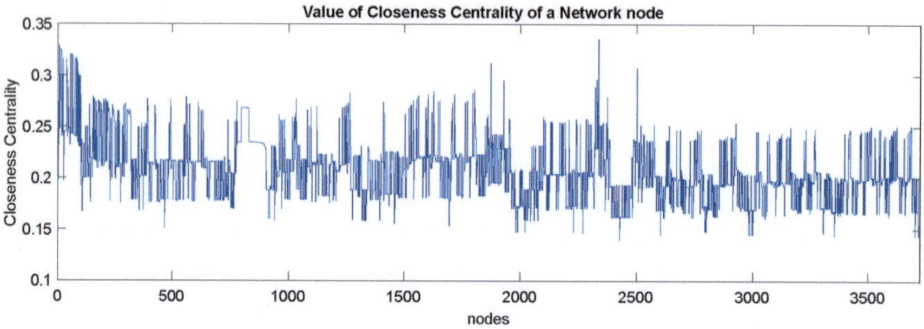

Fig. 4.48 Closeness centrality of network of water temperature

```
%Calculation of network Betweenness Centrality

% Net_Water_temp is the name of adjacency matrix that we have
                                          calculated previous

>> Between_Water_temp= betweenness_centrality_network
(Net_Water_temp);


%   Input:       adj       adjacency matrix
%
%   Output:      Between_Water_temp       Betweenness Centrality
vector per node

%                mean of clustering coefficient

Between_Water_temp =0.0011
```

To calculate the closeness centrality, we write in Matlab, while in Fig. 4.48, we observe the values of centrality of each node.

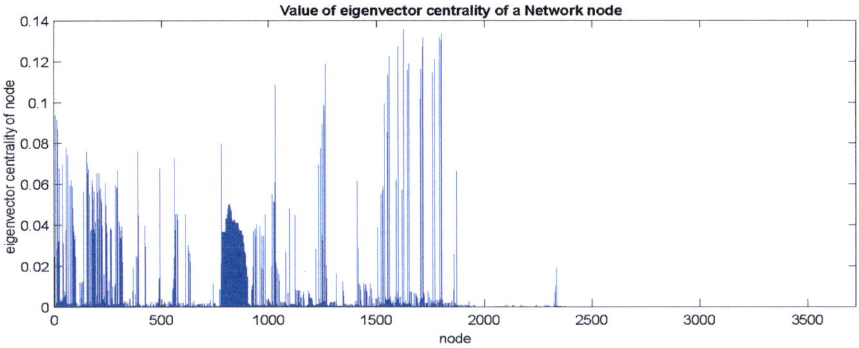

Fig. 4.49 Eigenvector centrality of network of water temperature

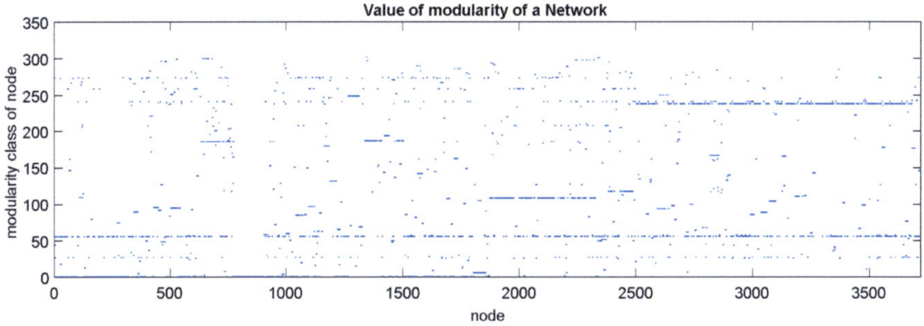

Fig. 4.50 Modularity class of each node of the network of water temperature

```
%Calculation of network closeness Centrality

% Net_Water_temp is the name of adjacency matrix that we have
                                         calculated previous

>> Between_Water_temp= closeness_centrality_network
(Net_Water_temp);


%   Input:      adj      adjacency matrix
%
%   Output:     Between_Water_temp      Betweenness Centrality
vector per node

%               mean of clustering coefficient

Between_Water_temp =0.0011
```

Figures 4.49 and 4.50 illustrate the results of the calculation of eigenvector centrality and modularity, respectively, for the case of the water temperature network.

4.5.2 *Examples of Simulation Data (Magnetohydrodynamics Time Series)*

In this section, an example of the application of the methodology for transforming a time series into a network, in a time series derived from simulation, will be presented. In particular, velocity time series of hydrodynamic and magnetohydrodynamic (MHD) turbulent flow are analyzed. The main scope is to understand the mechanism of fluid patterns modification due to the external magnetic field. The time series used was extracted from direct numerical method simulation, consisting of 3600 values.

In order to be able to complete all the steps of the analysis, below is the complete script file from plotting the data to converting to a graph and calculating the topological measures of the network.

```
%Script file example of time series to network

% %  Analysis of magnetohydradynamic time series

%    m_mu1 the name of the times series

%% plotting

figure (1)

plot(m_mu1)
title ('Time series','fontSize',11)
xlim([0 3601])
xlabel('time')
ylabel('Velocity')

figure (2)

DM_mu1=diff(m_mu1)
plot(DM_mu1)
xlim([0 3601])
title ('First Difference Time series','fontSize',11)
xlabel('time')
ylabel('Velocity')

%% convert to network

Net_x_nmu1=visibilitynet(DM_mu1);


%% plot adj
% Net_x_nmu1 the name of the network

figure ()

spy(Net_x_nmu1)
title ('Adjacency matrix','fontSize',11)
print -DMeta


%% convert to GEPHI file

 writetoPAJ(Net_x_nmu1, 'Net_x_adnmu1', 0)
```

The results of the analysis are presented in the following figures. The first figure shows the time series of speed, while the second is the time series of the first differences (Figs. 4.51 and 4.52).

Figure 4.53 illustrates the connectivity table of the network nodes. We can see that there are small–large squares which are connected in a range of points (≈800, ≈2100, ≈3000).

In Fig. 4.54, the corresponding network is shown, using the degree measure for illustration.

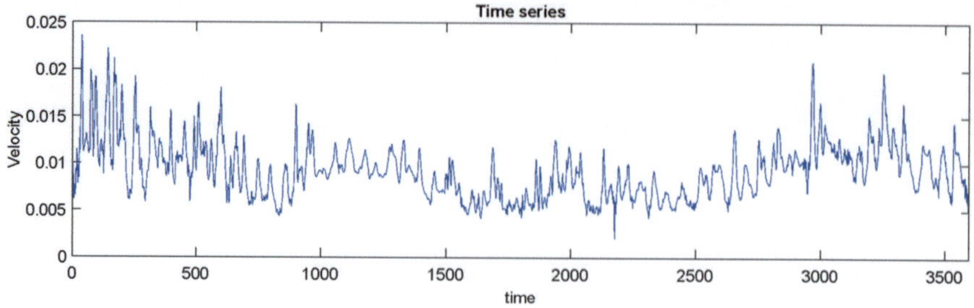

Fig. 4.51 Time series of velocity

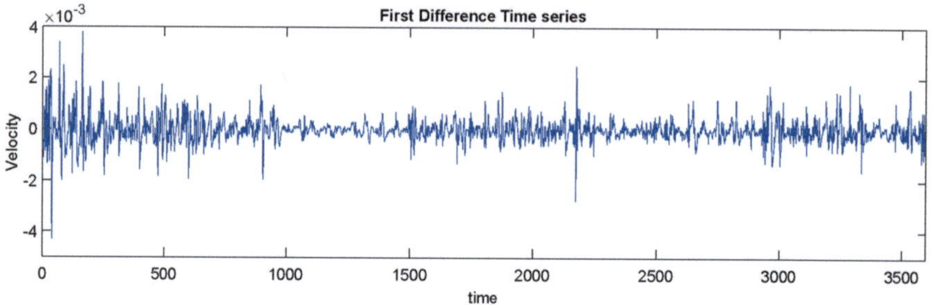

Fig. 4.52 First difference of time series of Fig. 4.51

Fig. 4.53 Adjacency matrix of the corresponding network

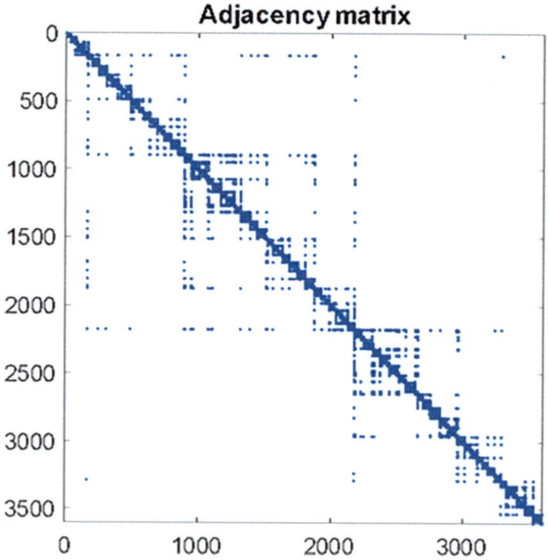

Fig. 4.54 Network of magnetohydrodynamic time series. The most intense color and largest nodes are the nodes that have the highest degree

The network of Fig. 4.54 is plotted based on the property of the degree of the nodes of the network; there is a color scale, and the greater the degree of the node, the larger the size of the corresponding marker. In this way, we can easily locate the nodes and then the corresponding values of the time series, since as we have mentioned there is a correspondence in the numbering of the values of the time series with that of the nodes of the network.

In this case, we can identify the points of the time series where they have a higher value/fluctuation than the rest. In general, in this way we identify the points of change of the dynamic state in the evolution of the time series. In the next figure, we present only the nodes with the maximum value of node degree measure. From Fig. 4.55, we can identify the points of the time series where there is a change in the dynamic evolution of the time series.

In Fig. 4.56, the corresponding network is shown, using the modularity measure for illustration. We can distinguish four groups–communities of nodes. These different communities can characterize different regions in the time series.

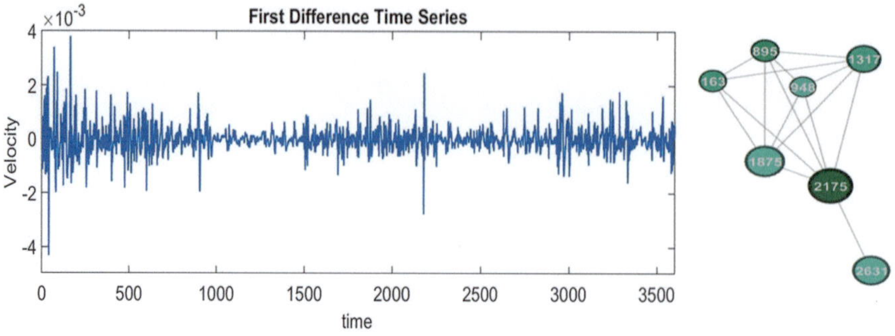

Fig. 4.55 Time series and part of the corresponding network based on the degree measure. The network shows the nodes with the highest values of the degree

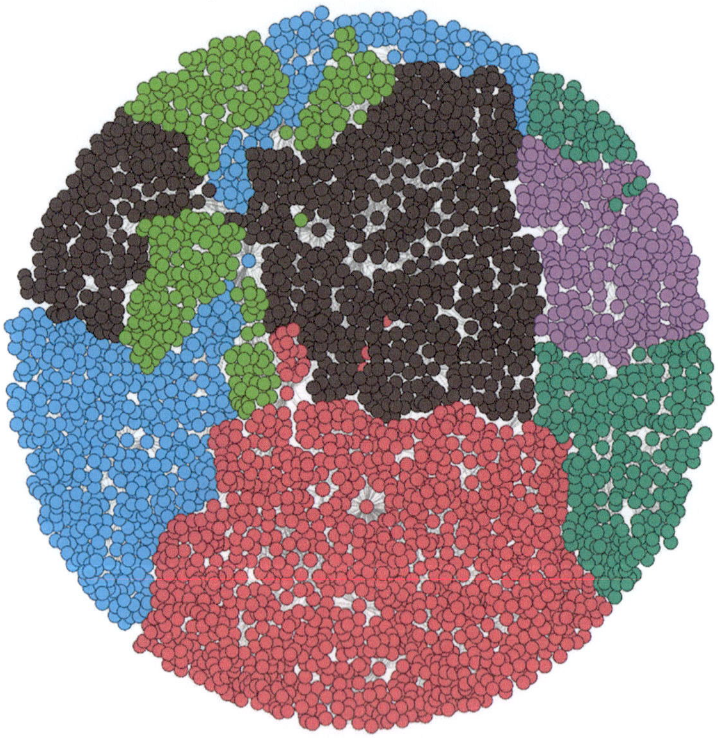

Fig. 4.56 Network from time series, based on the modularity measure

References

1. Albert, R., & Barabási, A. L. (2002). Statistical mechanics of complex networks. *Reviews of Modern Physics, 74*(1), 47.
2. Bastian, M., Heymann, S., & Jacomy, M. (2009). Gephi: An open source software for exploring and manipulating networks. *ICWSM, 8*, 361–362.

3. Boccaletti, S., Latora, V., Moreno, Y., Chavez, M., & Hwang, D. U. (2006). Complex networks: Structure and dynamics. *Physics Reports, 424*(4–5), 175–308.
4. Charakopoulos, A. K., Karakasidis, T. E., Papanicolaou, P. N., & Liakopoulos, A. (2014). The application of complex network time series analysis in turbulent heated jets. *Chaos: An Interdisciplinary Journal of Nonlinear Science, 24*(2).
5. Donges, J. F., Zou, Y., Marwan, N., & Kurths, J. (2009). Complex networks in climate dynamics. *The European Physical Journal Special Topics, 174*(1), 157–179.
6. Donner, R., & Donges, J. (2012). Visibility graph analysis of geophysical time series: Potentials and possible pitfalls. *Acta Geophysica, 60*(3), 589–623.
7. Donner, R. V., Small, M., Donges, J. F., Marwan, N., Zou, Y., Xiang, R., & Kurths, J. (2011). Recurrence-based time series analysis by means of complex network methods. *International Journal of Bifurcation and Chaos, 21*(04), 1019–1046.
8. Erdős, P., & Rényi, A. (1959). On random graphs I. *Publicationes Mathematicae Debrecen, 6*, 290–297.
9. Lacasa, L., Luque, B., Ballesteros, F., Luque, J., & Nuno, J. C. (2008). From time series to complex networks: The visibility graph. *Proceedings of the National Academy of Sciences, 105*(13), 4972–4975.
10. Lacasa, L., Luque, B., Luque, J., & Nuno, J. C. (2009). The visibility graph: A new method for estimating the Hurst exponent of fractional Brownian motion. *EPL (Europhysics Letters), 86*(3), 30001.
11. Zhang, J., & Small, M. (2006). Complex network from pseudoperiodic time series: Topology versus dynamics. *Physical Review Letters, 96*(23), 238701.
12. Small, M., Zhang, J., & Xu, X. (2009). Transforming time series into complex networks. In *Complex sciences* (pp. 2078–2089). Springer.
13. Watts, D. J., & Strogatz, S. H. (1998). Collective dynamics of 'small-world' networks. *Nature, 393*(6684), 440–442.

Chapter 5
Extended Case Studies

In this chapter, extensive examples of time series analysis are presented, based on the theory covered in previous chapters. The purpose of this chapter is to help the reader understand the theoretical concepts while simultaneously being able to reproduce the presented results by applying the routines provided in the book. In this way, we believe that the reader will be capable of applying these routines to other datasets as well. Example 1 includes field data, Example 2 used experimental data, while Example 3 performed simulated data.

5.1 Example 1: "Detection of Low-dimensional Chaos in Wind Time Series"

In this extended example, we explored the presence of low-dimensional deterministic chaos in wind time series obtained from a meteorological station [5]. Initially, we utilized techniques such as power spectrum and average mutual information function to extract characteristic times. Our examination of correlation dimension suggested the potential existence of a low-dimensional attractor, which provided significant evidence supporting the existence of low-dimensional chaotic dynamics within the wind time series.

The series consists of 30 years of weekly observations, provided by a meteorological station of horizontal wind speed measured as a weekly mean average, including all directions, resulting in a total number of 1488 records. In Fig. 5.1, the weekly mean wind speed vs time is presented, using script 5.1.

T. Karakasidis, A. Charakopoulos, *Time Series and Networks Analysis*,
https://doi.org/10.1007/978-3-031-92628-0_5

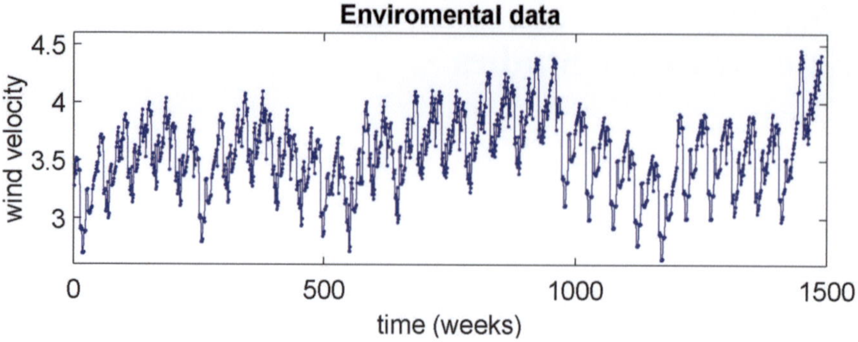

Fig. 5.1 Weakly mean wind speed (http://www.emy.gr/emy/el/)

```
%Script 5.1

plot(wind,'b.-','MarkerSize',10);    %'wind' the name of time
series

title('Environmental data','FontSize',20)
ylabel('wind velocity','FontSize',10)
xlabel('time (weeks)','FontSize',10)
```

We can obviously observe from the figure above the existence of periodicity.

To determine if the data contains trends, we apply the following script (5.2) (Fig. 5.2).

```
% Script 5.2
% Test for trend using the mean value

TS=input('Give the time series_:');
Size_segment=input('Give the time series (segment) length_:');
%50
Overlap=input('Give the overlap of segments_');

[TS_segments,index,reject] =slideWindow(TS, Size_segment, Ovelap);
TS_segments(TS_segments==0)=NaN;
columnMeans = mean(TS_segments,'omitnan');
plot(columnMeans,'o')
xlabel('Points');
ylabel('mean');
```

As we can see, using linear regression, the dotted line is created which indicates the trend. Then, the trend is removed using the first differences approach (script 5.3), and the corresponding time series appear in Fig. 5.3.

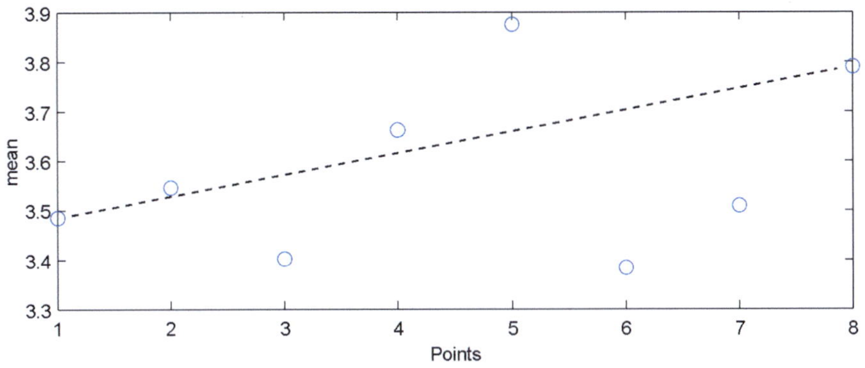

Fig. 5.2 Successive segments mean the wind speed time series

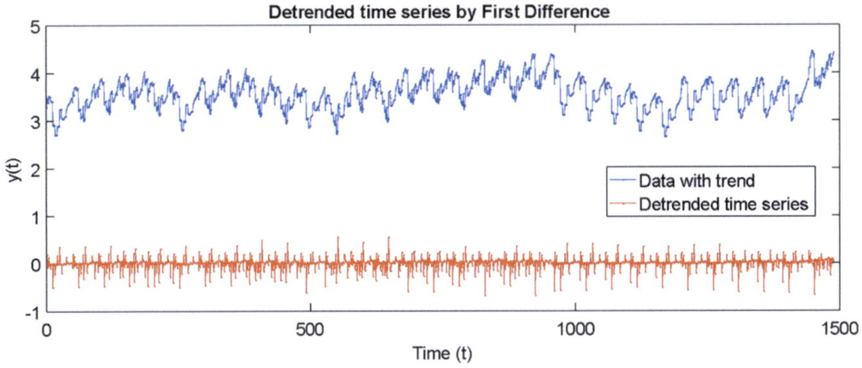

Fig. 5.3 Original and detrended time series using first difference

```
% Script 5.3
% Detrend data with applying first difference

TStrend = input('Give the time series with trend_:');
length = input('Give the time series length_:');
t=1:length;
Diff_detreded=diff(TStrend)    % Diff_y1 the time series without
trend
plot(t,TStrend,'.-');
hold on
plot(Diff_detreded,'.-');
legend('Data with trend','Detrended time series')
xlabel('Time (t)');
ylabel('y(t)');
title('Detrended time series by First Difference','FontSize',14)
```

Then we employ the mutual information analysis in order to extract a time lag for the reconstruction of space phase following the script 5.4, and the results appear in

Fig. 5.4. As we can see the first minimum of the mutual information of the first differences occur at $\tau = 1$. For the calculation of the lag time, it is recommended to be done on the time series where the trend has been removed.

```
% Script 5.4
% Plot time series and compute mutualinformation

data=input('Time series name_');
tmax=input('Time Lag_');
figure
subplot(2,1,1);
plot(data,'b.-','MarkerSize',6);
axis([0 1480 2 5])
title('Time series','FontSize',10)
ylabel('y(t)')
xlabel('(t) Time')
subplot(2,1,2);
[mutM] = mutualinformation_b(data, tmax)
```

In order to determine that it is necessary to remove the trend, below is given the initial time series as well as the result of finding the lag time using mutual information function. We can see from Fig. 5.5 that there is a smooth decline without showing a sharp plunge in values.

Next we employ the false nearest neighbor method in order to select the minimal embedding dimension. This method is based on the assumption that two points that are near to each other in the sufficient embedding dimension should remain close as the dimension increases. We obtain the results appearing in Fig. 5.6 using the script 5.5, where tau = 1 and nmax = 10.

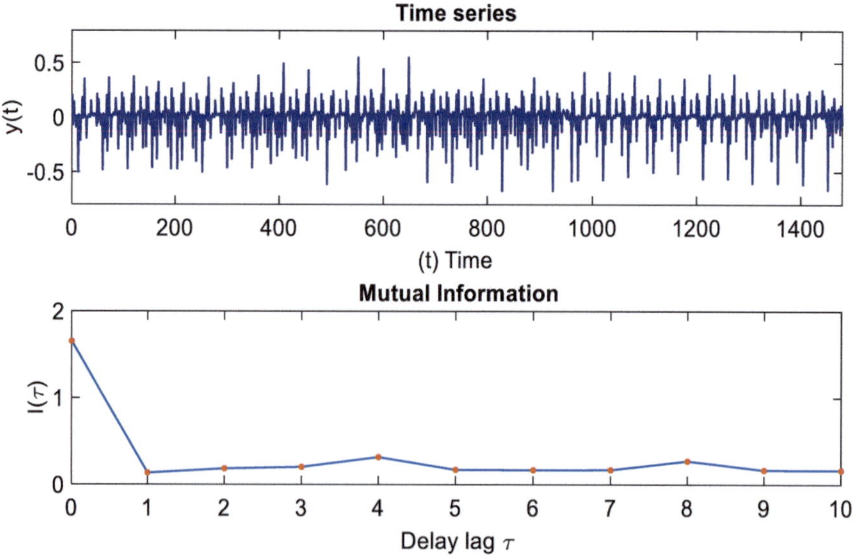

Fig. 5.4 Results of mutual information function of detrended time series

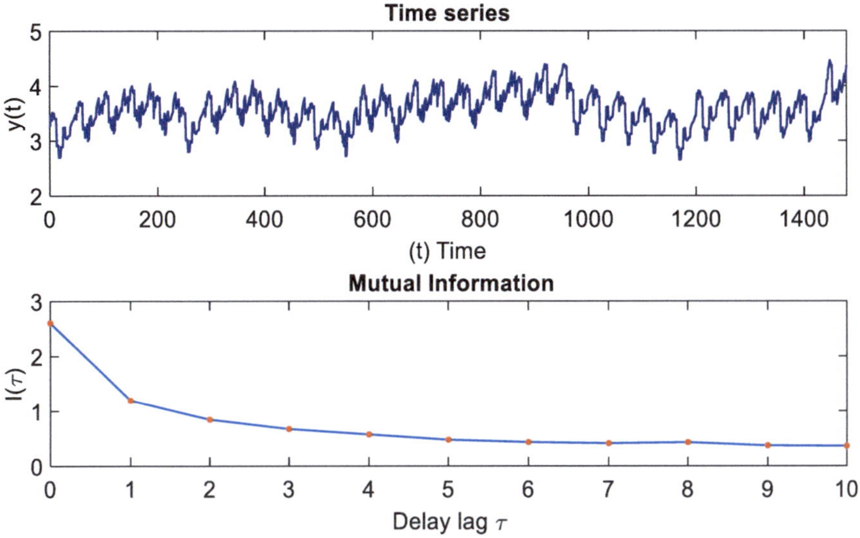

Fig. 5.5 Results of mutual information function of initial time series

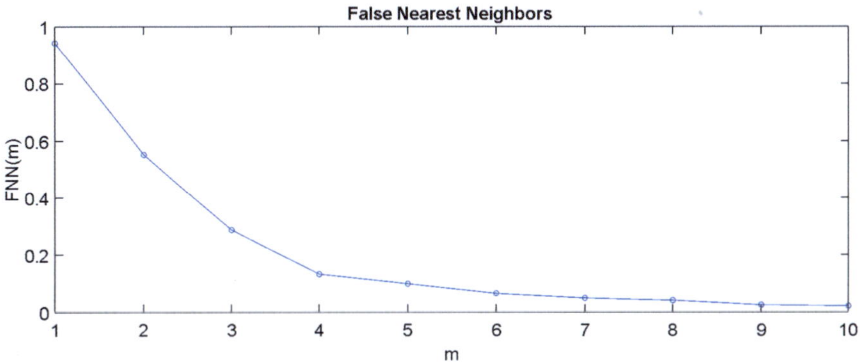

Fig. 5.6 False nearest neighbor function with time embedding 10

```
% script 5.5
% False Nearest Neighbors
% calculate the embedding dimension

function FnM = falsenearest(wind,1,10)
% FnM = falsenearest(xV,tau,mmax,escape,theiler,tittxt)
% Computes the false nearest neighbors starting from 1 to 'mmax'
%embedding dimensions.


% INPUT
%  xT       : time series
%  tau      : delay time. If empty, tau=1
%  mmax     : maximum embedding dimension.
%
```

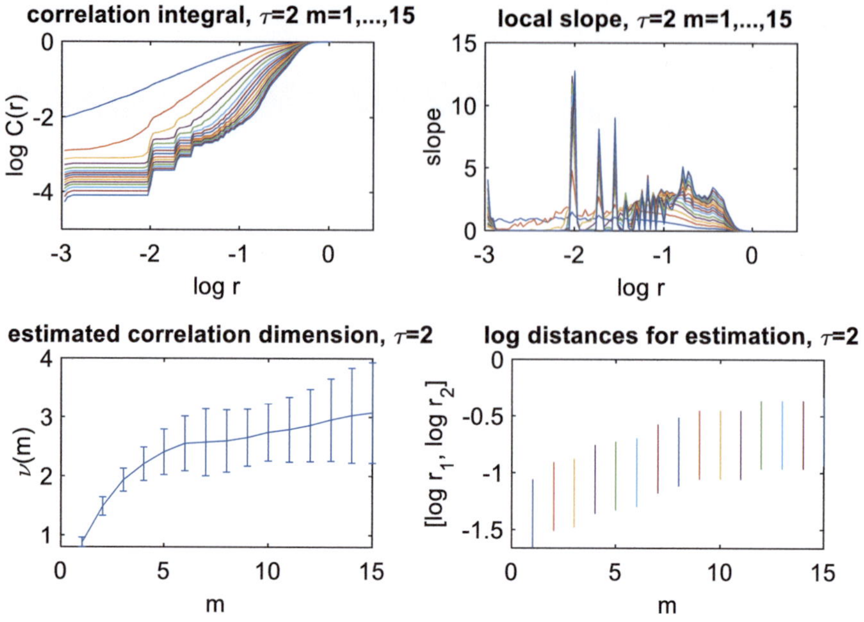

Fig. 5.7 Estimation of correlation dimension

Then we evaluate the correlation dimension, using the code presented in previous subchapter, which is an indicator of existing low-dimensional chaos (Fig. 5.7).

From the above figures, we can see that the time series of wind speed presents low-dimensional chaos, with a correlation dimension between 2 and 3.

5.2 Example 2: "Identification of Spatiotemporal Phenomena Using Non-linear Time Series Analysis and Network Analysis Methods"

In this example, we describe a methodology for analyzing a spatiotemporal problem and specifically the identification of regions in a liquid turbulent flow problem in an experimental arrangement [2–4].

Discriminating the state of the fluid region as a function of the distance from the jet axis is crucial challenge. We provide a methodology for studying turbulent flows, specifically for identifying different regions of the jet (from the point of view of their dynamical behavior) through nonlinear methodologies and also employed the complex network analysis. In order to understand the example, first we present the basic elements of the problem, which are common to both approaches to analysis, and then we focus on the two methods separately.

The data originate from temperature recordings in a vertical turbulent heated jet where time series were recorded along a horizontal cut through the jet axis by transforming them to complex networks. The time series are transformed into complex networks using the visibility graph method, and then for each network, we evaluated the main topological network properties to demonstrate how they can effectively distinguish different dynamical regimes of the liquid regions.

A schematic flow representation of the turbulent jet is sketched in Fig. 5.8, where in the case of fully developed turbulence, three different regions are expected:

(a) The boundary region (BR) located at large distances from the jet axis (boundary with ambient fluid).
(b) The inner region (IR) positioned between the boundary region and the center of the jet.
(c) The jet axis region (JR) close to the jet axis.

In this study, 21 recordings of temperature time series obtained at various measurement locations along a horizontal cut of the flow were employed. The sampling period at each location was 40 s at a frequency of 60 Hz. The first time series was recorded at position $x = 32.40$ cm (the first at top) and the last one at $x = 46.50$ cm (the last at the bottom), and the corresponding time series are displayed in Fig. 5.9. On the horizontal axis, time (t) is presented, while the vertical axis refers to each time series which is located as we move from the left boundary (position $x = 32.40$ cm) of the tank to the right (position $x = 46.50$ cm) (Fig. 5.10).

Fig. 5.8 A schematic turbulent jet flow

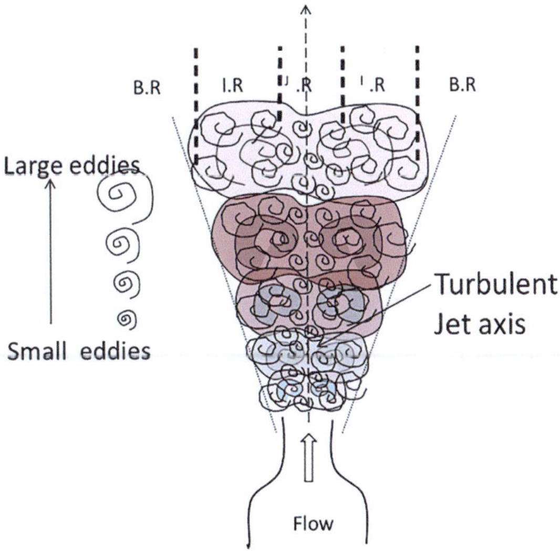

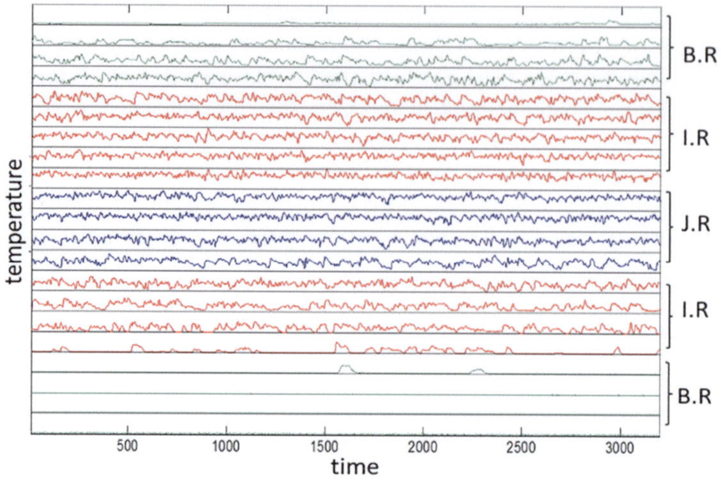

Fig. 5.9 The time series recorded at various measurement distances from the jet axis. The green time series correspond to the boundary region, while the blue and red time series represent the jet axis region and inner regions, respectively

Fig. 5.10 A shadowgraph view of the experimental setup. The green areas indicate the boundary region, while the blue and red areas represent the jet axis region and inner regions, respectively

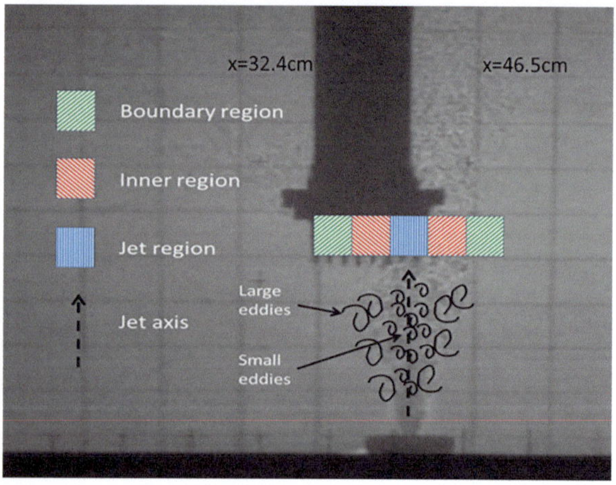

5.2.1 *Nonlinear Analysis*

In this approach, for each temperature time series measured at different locations, we estimated mainly nonlinear measures such as mutual information combined with descriptive statistics measures, as well as some linear and nonlinear dynamic detectors such as Hurst exponent, detrended fluctuation analysis (DFA), and Hjorth parameters.

First for each time series, we evaluate the Hurst exponent, using the following command in Matlab (Fig. 5.11).

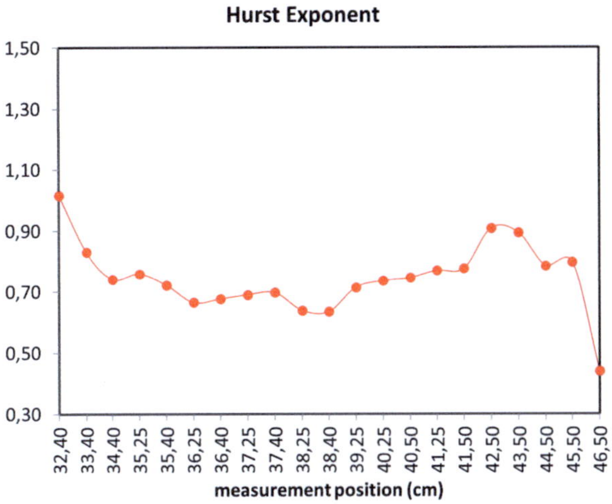

Fig. 5.11 Hurst exponent of the time series as a function of measurement positions with R/S method

```
%%
%Hurst Exponent

H= HurstExponent(XV)% XV is the name of each time series
```

The Hurst exponents exhibit their lowest values within the range of 36.25–38.40 cm, corresponding to the jet axis region, thereby distinguishing these time series and their respective measurement regions from others. Generally, the Hurst exponent values exceed 0.5, except at $x = 46.5$, which lies outside the jet region in the ambient water region, reflecting a persistent behavior.

Figure 5.12 presents the **Hjorth parameters**, mobility, and complexity, of the time series. Notably, the highest mobility values are observed in the time series originating from the jet core, where the complexity values are at their lowest. In order to calculate the Hjorth parameters, we write the following in Matlab.

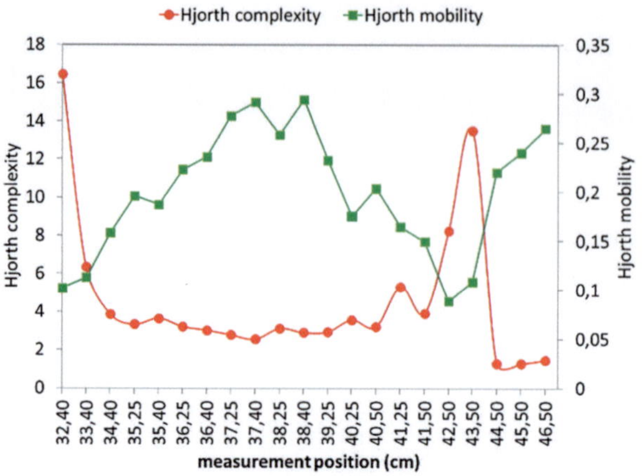

Fig. 5.12 Hjorth parameters mobility and complexity of the time series as a function of measurement positions

```
%Script 5.6
% Mutual Information of a matrix of time series

function [mutM_all] = mutualInformationMatrixavr(timeSeriesMatrix,
tmax, partitions, tittxt, type)
% MUTUALINFORMATIONMATRIXSINGLEPLOT computes and plots mutual
information for
% each column of a matrix of time series, with all results shown
in one figure.
% INPUTS:
%  timeSeriesMatrix : Matrix where each column represents a time
series
%  tmax             : Largest delay time to compute mutual
information

% OUTPUT:
%  mutM_all         : A cell array where each element contains the
mutual information for one time series
'b');

    % Validate input
    [n, FGMSeries] = size(timeSeriesMatrix);
    if nargin < 3 || isempty(partitions)
        partitions = ceil(sqrt(n / 5));
    end
    if nargin < 4
        tittxt = '';
    end
    if nargin < 5
        type = 'b';
    end

    % Initialize results
    mutM_all = cell(FGMSeries, 1);

    % Prepare figure
    figure;
    hold on;

    % Loop through each time series (column)
    for col = 1:FGMSeries
        xV = timeSeriesMatrix(:, col);
        mutM = computeMutualInformation(xV, tmax, partitions);
        mutM_all{col} = mutM; % Store results

        % Plot results
        if type == 'd'
            plot(mutM(:, 1), mutM(:, 2), '.', 'DisplayName',
['Series ', FGM2str(col)]);
        elseif type == 'c'
            plot(mutM(:, 1), mutM(:, 2), '-', 'DisplayName',
['Series ', FGM2str(col)]);
        else
            plot(mutM(:, 1), mutM(:, 2), '-o', 'MarkerSize', 6,
'DisplayName', ['Series ', FGM2str(col)]);
        end
    end

    % Customize plot
    grid on;
    xlabel('Delay (lag)', 'FontSize', 10, 'FontWeight', 'bold');
    ylabel('Mutual Information', 'FontSize', 10, 'FontWeight',
```

```
'bold');
     title(tittxt, 'FontSize', 12, 'FontWeight', 'bold');
     legend('show', 'Location', 'Best');
     hold off;
 end

function [mutM] = computeMutualInformation(xV, tmax, partitions)
% Computes mutual information for a single time series
    n = length(xV);
    h1V = zeros(partitions, 1); % Marginal for x(t+tau)
    h2V = zeros(partitions, 1); % Marginal for x(t)
    h12M = zeros(partitions, partitions); % Joint probabilities

    % Normalize data
    xmin = min(xV);
    xmax = max(xV);
    xV = (xV - xmin) / (xmax - xmin + eps); % Avoid division by
zero

    % Assign to partitions
    arrayV = min(floor(xV * partitions) + 1, partitions);

    % Initialize mutual information results
    mutM = zeros(tmax + 1, 2);
    mutM(:, 1) = (0:tmax)'; % Lag values

    % Compute mutual information for each lag
    for tau = 0:tmax
        ntotal = n - tau;
        h12M(:) = 0; % Reset joint histogram
        for t = 1:ntotal
            h12M(arrayV(t + tau), arrayV(t)) = h12M(arrayV(t +
tau), arrayV(t)) + 1;
        end

        % Compute marginals
        h1V = sum(h12M, 2); % Sum over rows
        h2V = sum(h12M, 1); % Sum over columns

        % Normalize probabilities
        h12M = h12M / ntotal;
        h1V = h1V / ntotal;
        h2V = h2V / ntotal;

        % Calculate mutual information
        mutS = 0;
        for i = 1:partitions
            for j = 1:partitions
                if h12M(i, j) > 0
                    mutS = mutS + h12M(i, j) * log(h12M(i, j) /
(h1V(i) * h2V(j) + eps));
                end
            end
        end
        mutM(tau + 1, 2) = mutS;
    end
end
```

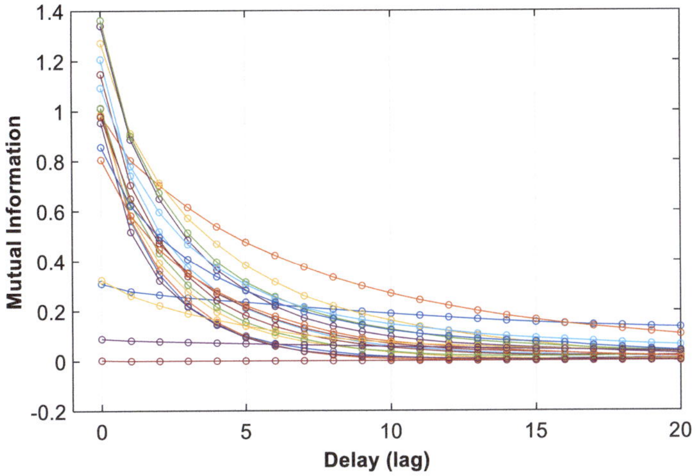

Fig. 5.13 Mutual information of the time series as a function of measurement positions

Using the following command in Matlab, we calculate the mutual information for each time series plotted in the same figure, giving as input the matrix with the time series (Fig. 5.13).

```
% Suppose the name of time series is mag_2

% Plot time series

>> plot(mag_2)
```

It can be observed that time series coming from different positions on the horizontal axis of the experiment show a different distribution of the values of the mutual information function. In this way, we can separate the time series in relation to their recording location.

Using the previously methods, it is evident that the jet regions can be distinguished through the analysis of the corresponding time series. Additionally, clustering technique can be applied in an effort to better discriminate the various regions of the jet as well as locate the jet axis. The hierarchy built by the clustering algorithm from each time series is represented by the dendrograms given in Fig. 5.14.

The horizontal axis represents each time series, while the vertical axis indicates the distance. At the top of the dendrogram, the position of each time series measurement along the horizontal axis is schematically displayed.

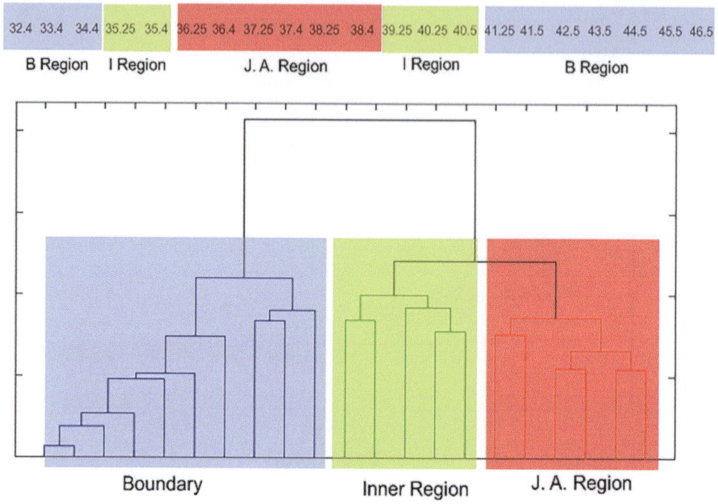

Fig. 5.14 Dendrogram based on the time series. The sketch above is a schematic representation of the location of the various regions in the measurement setup

5.2.2 *Complex Network Analysis*

In this approach, the main idea is to analyze and investigate temperature fluctuations from a vertical turbulent heated jet where temperature time series were recorded along a horizontal cut through the jet axis by transforming them to complex networks [1].

We convert each time series to a network and then we calculate the topological measures.

In Fig. 5.15, the diameter, modularity, and clustering coefficient are presented as functions of the horizontal position along the reconstructed visibility algorithm, presented in Chap. 4. The horizontal axis represents the measurement locations of each time series. The diameter profile using the visibility algorithm exhibits its lowest value at $x = 37.40$ cm. Notably, measurements on the far right and far left correspond to ambient water, representing a distinct dynamical regime compared to the flow region, which spans approximately from 35 to 42 cm. Therefore, when referencing minimum or maximum values, we focus on the flow region, excluding the full extent of the measurement data.

From the modularity results, it is observed that the lowest modularity values within the flow region occur for the time series at $x = 37.40$ cm. Generally, higher modularity values imply fewer communities, while the lowest modularity value at $x = 37.40$ cm indicates a network with many smaller communities. This behavior is linked to the underlying physics: near the jet axis, the dynamics are influenced by small, short-lived vortices that frequently perturb the system, reducing connectivity between successive states. Conversely, closer to the boundaries, large, long-lived

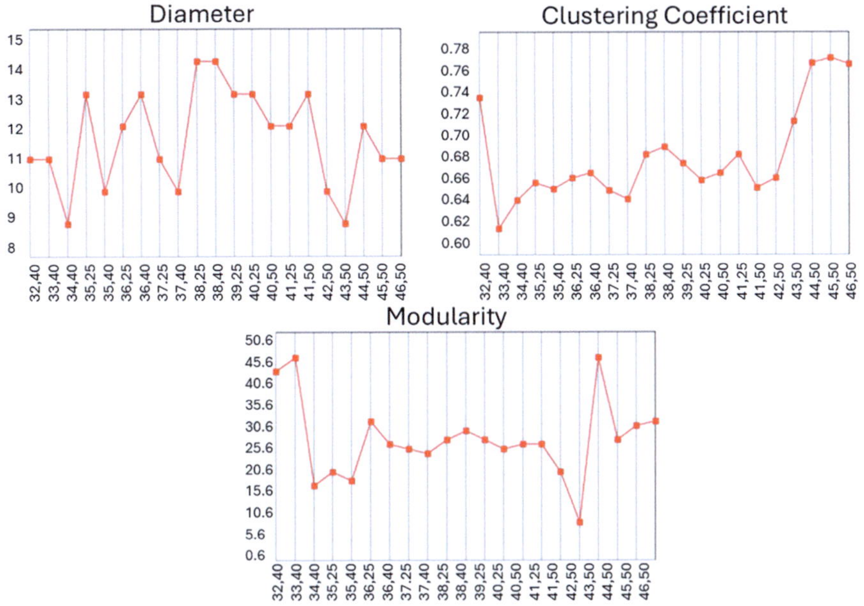

Fig. 5.15 Diameter, modularity, and clustering coefficient of the networks along the horizontal axis

structures dominate, leading to more connected states and fewer, larger communities.

In the inner region (x = 36.40 cm to x = 40.25 cm), the clustering coefficient reaches its lowest value at x = 37.40 cm. The clustering coefficient reflects the probability that the neighbors of a node are connected, indicating the likelihood that a node's "friends" are also "friends" with each other. For the network derived from the time series at x = 37.40 cm, nodes exhibit greater independence compared to networks constructed from other time series in the flow region. This behavior again reflects system dynamics, as the small, short-lived vortices near the jet axis lead to frequent disruptions, resulting in faster memory loss in system evolution and fewer connected neighboring points (nodes) in the network. Moving toward the boundaries, the presence of long-lived structures promotes longer memory effects, resulting in greater connectivity among network nodes.

In summary, the lowest values of diameter, average path length, modularity, and clustering coefficient are observed for the time series at x = 37.40 cm. The interpretation of these topological properties identifies this position as corresponding to the jet axis. Interestingly, conventional hydrodynamic methods, using exponential fitting, estimate the jet axis position at x = 37.75 cm. More broadly, the spatial variation in network topological properties enables clear differentiation between the jet axis region and other parts of the jet.

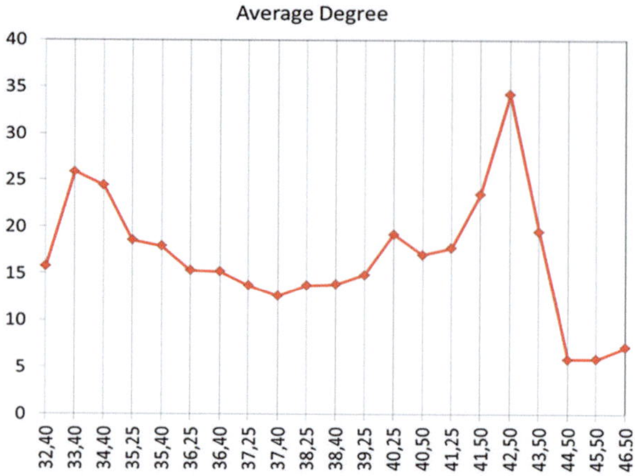

Fig. 5.16 Average degree of the networks along the horizontal axis

The average degree profile of each network constructed by visibility graph is displayed in Fig. 5.16. The average degree of a network is the average of the degrees over all nodes in the network.

The profile of the average degree is quite symmetric. Near the boundary (except for the location $x = 32.40$ cm on the left and the region $x = 42.50$–46.50 cm on the right, which lies outside the increased turbulence region and in fact corresponds to ambient water) degree presents high values. As we move away from the boundary toward the inner region, the degree value decreases and obtains its lowest value at the position $x = 37.40$ cm.

Another noteworthy aspect is that the degree distribution of a network is considered one of the most significant properties of a network. According to Lacassa et al., the visibility graph network of a time series has a power law degree distribution $P(k) = k^{-\gamma}$ and is characterized as a scale-free network. Figure 5.17 presents the degree distribution of the networks constructed from selected time series recorded in the three different regions of the flow as discussed already: one from the boundary region, one from inner region, and one from the jet region.

The results show that the networks follow a power law tail distribution, $P(k) = k^{-\gamma}$, for $k > 10$ with varying power exponents γ. This suggests that the networks are scale-free in this range and may exhibit fractal characteristics. The number of high-degree nodes ($k > 70$) is limited to one or none, so they are excluded from the slope calculation. Notably, the exponent γ decreases as we move from the jet region to the boundary, indicating a variation in slopes: $\gamma = 3.22$ in the jet region, $\gamma = 2.78$ in the inner region, and $\gamma = 1.68$ at the boundary. The jet region exhibits the steepest slope among the three.

As we move away from the boundary region toward the jet axis region, the presence of small short living structures is enhanced resulting in a more frequent disruption of the dynamics of the corresponding fluid regions which results in a faster loss of memory in the time series and as a result less nodes connected. Such events

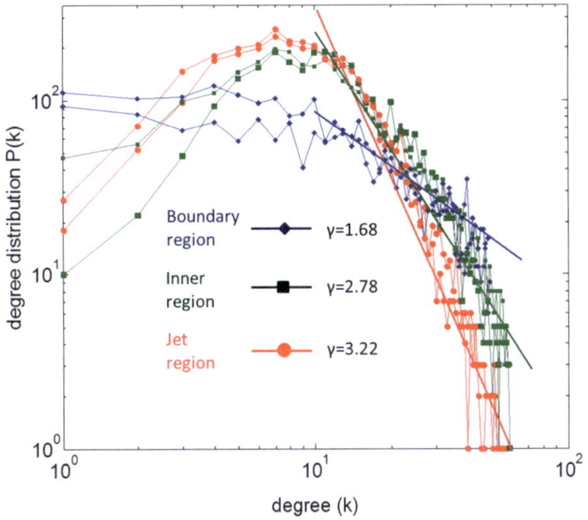

Fig. 5.17 Degree distribution of networks from time series near the boundary (blue), near the inner region (green), and close to jet region (red)

reduce significantly the number of nodes with high number of neighbors which are a result of the larger and longer living structures which dominate far from the jet axis. Notably, there is also a variation of the low neighbor nodes ($k < 10$) depending on the location of the measurement station.

5.3 Example 3: "Analysis of Magneto-hydrodynamic Channel Flow Through Complex Network Analysis"

In this example, we analyze hydrodynamic (HD) and magnetohydrodynamic (MHD) velocity time series from direct numerical simulations (DNS) using the visibility graph method [2]. We investigate whether the flow can be classified into three distinct regions based on turbulent boundary-layer theory. Additionally, the study explores the identification of different dynamical regions and hidden characteristic patterns in the presence and absence of a magnetic field.

First, it is shown that the velocity time series recorded in different regions of the flow exhibits distinct topological network structures. Additionally, various topological properties of the resulting networks effectively capture and distinguish the three turbulent regions—viscous sublayer, buffer layer, and log-law layer—in both hydrodynamic and magnetohydrodynamic (MHD) flows.

Figure 5.18 illustrates a schematic representation of the flow field geometry in a turbulent channel flow, following the principles of turbulent boundary-layer theory. According to wall-bounded turbulence theory, such flows consist of three distinct regions. The region closest to the walls, known as the "Viscous Sublayer" (VS), is a

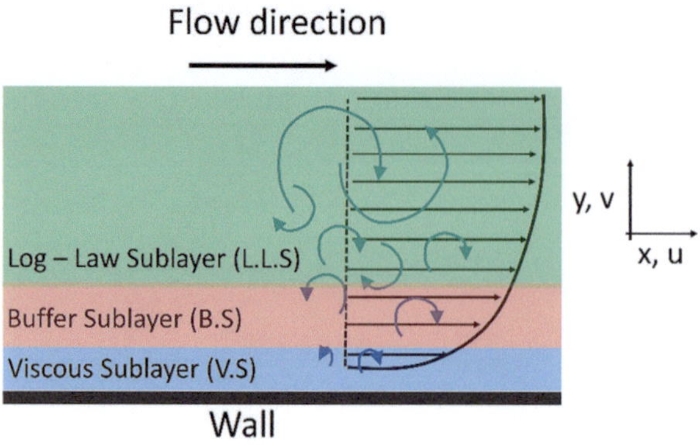

Fig. 5.18 Schematic flow of turbulent channel flow near a wall

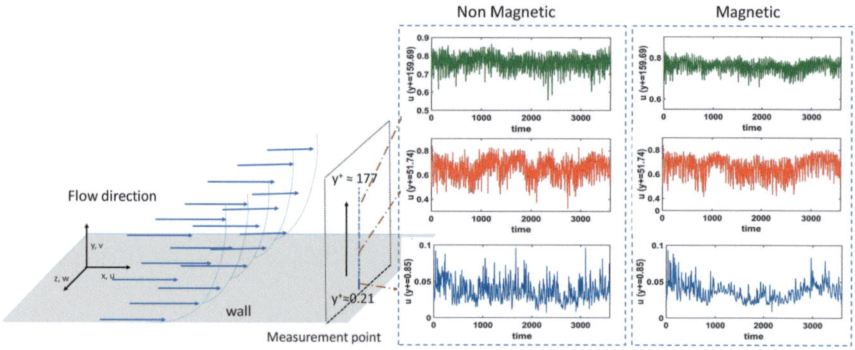

Fig. 5.19 A schematic of the 3D channel flow simulation setup, along with instantaneous streamwise velocity profiles at three different wall-normal heights, is shown

thin layer where viscous effects dominate. Above this lies the "Buffer Sublayer" (BS), characterized by a high time-averaged velocity and fully developed turbulence. The outermost region, referred to as the "Log-Law Sublayer" (LLS), is dominated by turbulent shear stress, with turbulence exhibiting significantly larger flow structures compared to those near the boundary (Fig. 5.19).

The left panel represents the case without a magnetic field, while the right panel illustrates the flow behavior under the influence of a magnetic field. The blue time series refer to the VS region, while red and green time series refer to the BS region and LLS region, respectively.

A closer examination of the time series reveals that, in the presence of a magnetic field, the fluctuations are significantly reduced and appear smoothed out. Table 5.1 provides a summary of the expected regions and the approximate measurement positions along the wall-normal direction (Fig. 5.20).

Table 5.1 Expected flow regions

Region name	Measurement position along wall-normal direction (approximately)	Description
Viscous sublayer (VS)	From $y^+ = 0.21$ to $y^+ = 6.39$	Area close to the channel wall
Buffer sublayer (BS)	From $y^+ = 7.60$ to $y^+ = 29.76$	Region next to the boundary
Log-law sublayer (LLS)	From $y^+ = 34.75$ to $y^+ = 177.11$	The area at higher distance from the channel wall

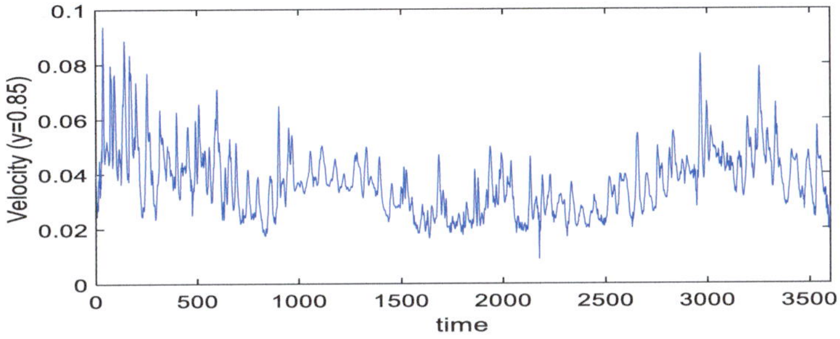

Fig. 5.20 Time series measurement at $y^+ = 0.85$

```
script 5.7
% Suppose the name of time series is mag_2
>> Net_Mag_2=visibilitynet(mag_2)
```

Each time series was first transformed into networks using the visibility graph algorithm, as previously described in detail. To transform the time series into a network, we use the following script (5.) in the Matlab. Figure 5.21 shows the network connectivity (adjacency matrix) using script 5.7.

```
script 5.8
% Suppose the name of time series is mag_2
% Net_Mag_2 is the name of network
>> spy(Net_Mag_2)
```

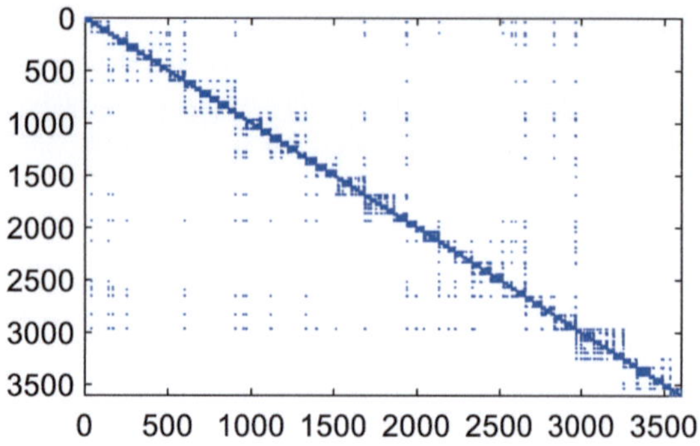

Fig. 5.21 Adjacency matrix of time series measurement at $y^+ = 0.85$

```
script 5.9

% Net_Mag_2 is the name of network

% Transform network to net format

% NET_MAG_2 the name of net format file

>> writetoPAJ(Net_Mag_2,'NET_MAG_2',0)
```

In order to be able to visualize the network in the Gephi software, we use the following script 5.8. Figure 5.22 shows the network layout.

```
script 5.10

%Calculation of network degree of each node

% Net_Mag_2 is the name of adjacency matrix that we have
calculated previous

% D_Net_Mag_2 the degree  of network

>>D_Net_Mag_2 =Degree_network(Net_Mag_2)
```

Subsequently, we analyzed the topological properties of the resulting complex networks using the following script files. The computed metrics, degree, closeness centrality, clustering coefficient, eigenvector centrality, and betweenness centrality are presented as functions of the measurement position along the wall-normal direction in inner units (y^+). These results are illustrated in Fig. 5.23a–e, where dashed lines indicate the boundaries of different flow regions.

Fig. 5.22 Network layout of time series measurement at $y^+ = 0.85$

```
script 5.11

%Calculation of network Betweenness Centrality

% Net_Mag_2 is the name of adjacency matrix that we have calculated
                                                          previous

>> Between_Net_Mag_2= betweenness_centrality_network(Net_Mag_2);


%   Input:      adj      adjacency matrix
%
%   Output:     Between_Net_Mag_2 Betweenness Centrality vector
per node

%               mean of clustering coefficient
```

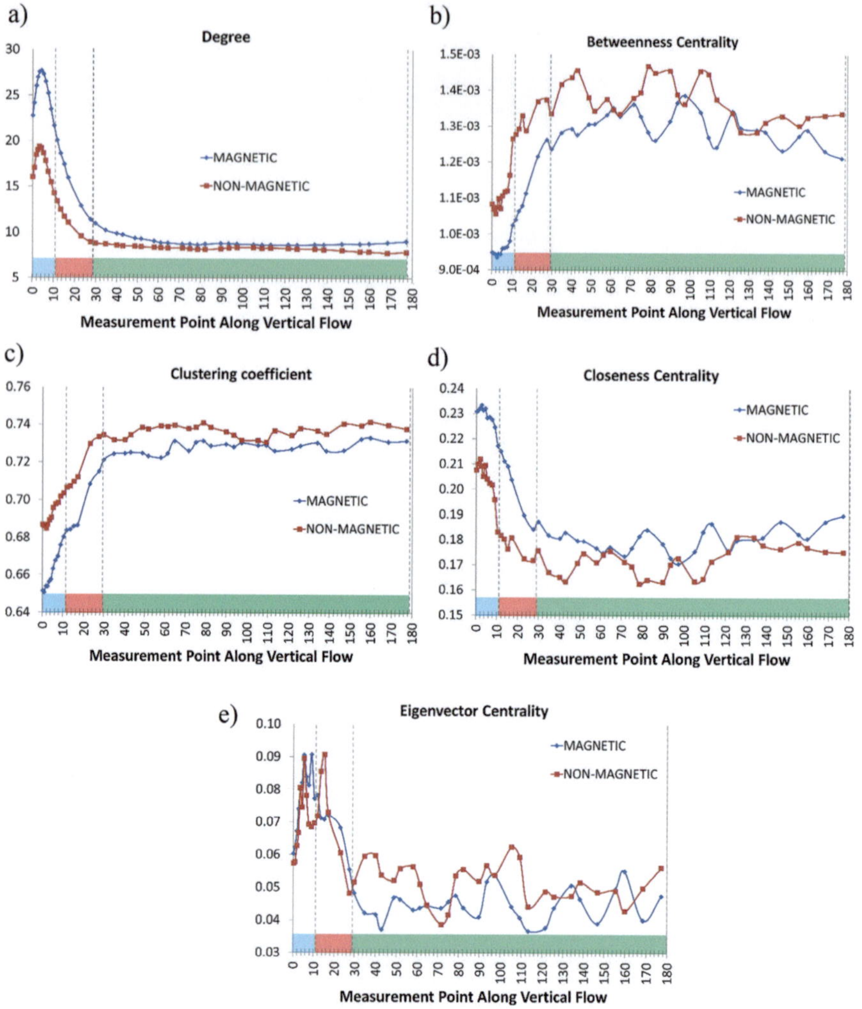

Fig. 5.23 (**a**–**e**) Topological properties of the network derived from time series measurement at $y^{+} = 0.85$

```
script 5.12

%Calculation of network clustering coefficient

% Net_Mag_2 is the name of adjacency matrix that we have calculated
                                                        previous

>> [Cl, avCl_ Net_Mag_2]=clustering_coef_network(Net_Mag_2);


%   The clustering coefficient is the fraction of triangles around
a node
```

```
%    Input:      adj        adjacency matrix
%
%    Output:      Cl        clustering coefficient vector per node

%                avCl_Net_Mag_2 mean of clustering coefficient
```

```
script 5.13

%Calculation of network closeness Centrality

% Net_Mag_2 is the name of adjacency matrix that we have calculated
                                                       previous

>> Clossenes_Net_Mag_2= closeness_centrality_network (Net_Mag_2);


%    Input:      adj        adjacency matrix
%
%    Output:     Clossenes_Net_Mag_2       Clossenes Centrality
vector per node

%                mean of clustering coefficient
```

```
script 5.14

%Calculation of network closeness Centrality

% Net_Mag_2 is the name of adjacency matrix that we have calculated
                                                       previous

>> Eignv_Net_Mag_2= eigenvector_centrality_network (Net_Mag_2);


%    Input:      adj        adjacency matrix
%
%    Output:     Eignvector_Net_Mag_2       eignvector Centrality
vector per node

%                mean of eignvector Centrality
```

Both in the absence and presence of a magnetic field, the network properties exhibit a similar profile but on different scales. A closer examination reveals that near the channel wall (viscous sublayer, VS), the degree value initially increases up to approximately $y^+ = 7$ (marked as a light blue area), then gradually decreases until around $y^+ = 30$ (buffer layer, BS), and remains nearly constant beyond this point in the log-law sublayer (LLS). This average degree profile aligns with the physical interpretation of the data and the network construction method, as time series from the buffer and log-law sublayers exhibit greater variability compared to those from the viscous sublayer.

References

1. Charakopoulos, A. K., Karakasidis, T. E., Papanicolaou, P. N., & Liakopoulos, A. (2014). The application of complex network time series analysis in turbulent heated jets. *Chaos: An Interdisciplinary Journal of Nonlinear Science, 24*(2).
2. Charakopoulos, A., Karakasidis, T., & Sarris, I. (2021). Analysis of magnetohydrodynamic channel flow through complex network analysis. *Chaos: An Interdisciplinary Journal of Nonlinear Science, 31*(4).
3. Charakopoulos, A. K., Karakasidis, T. E., Papanicolaou, P. N., & Liakopoulos, A. (2014). Nonlinear time series analysis and clustering for jet axis identification in vertical turbulent heated jets. *Physical Review E, 89*(3), 032913.
4. Charakopoulos, A. K., Karakasidis, T. E., & Liakopoulos, A. (2015). Spatiotemporal analysis of seawatch buoy meteorological observations. *Environmental Processes, 2*, 23–39.
5. Karakasidis, T. E., & Charakopoulos, A. (2009). Detection of low-dimensional chaos in wind time series. *Chaos, Solitons & Fractals, 41*(4), 1723–1732.

Index

T. Karakasidis, A. Charakopoulos, *Time Series and Networks Analysis*,
https://doi.org/10.1007/978-3-031-92628-0

MIX
Papier aus verantwortungsvollen Quellen
Paper from responsible sources
FSC® C105338

If you have any concerns about our products,
you can contact us on
ProductSafety@springernature.com

In case Publisher is established outside the EU,
the EU authorized representative is:
Springer Nature Customer Service Center GmbH
Europaplatz 3, 69115 Heidelberg, Germany

Printed by Libri Plureos GmbH
in Hamburg, Germany